Table of Contents

- ➤ MSOE is a leader in fluid power research and education for more than 50 years.
- ➤ MSOE provides training on campus, at customer sites and internationally.
- ➤ MSOE certifies all courses and eligible to grant Continuing Education Credit Units (CEU) for the participants of the professional education programs.
- ➤ MSOE is one of the most recognized institutions nationwide in terms of fluid power practical-oriented education.
- ➤ MSOE provides long term technical skills development solutions for industry.

Milwaukee School of Engineering:

MSOE is a private, non-profit university with about 2,600 students that was founded in 1903. MSOE offers bachelor's and master's degrees in engineering, business, mathematics and nursing, as well as professional education courses and certifications in fluid power. The university has a national academic reputation; longstanding ties to business and industry; and dedicated professors with real-world experience.

UNIVERSITY

Sheku Kamara is the dean of applied research at Milwaukee School of Engineering (MSOE) and oversees the activities of the Applied Technology Center™ (ATC). The ATC uses MSOE faculty, staff and student expertise to solve technological problems confronting business and industry. The center undertakes more than 250 industry-sponsored research projects that are focused on providing real solutions to some of industry's biggest problems every year. This work is completed through several centers of excellence within the ATC including the Center for BioMolecular Modeling (CBM), Fluid Power Institute™ (FPI), Professional Education Research & Development (PERD) and the Rapid Prototyping Center (RPC). Since 2004, he has been a technical advisor to the RAPID conference organized by SME and past chair of the Additive Manufacturing Users Group (AMUG). Kamara is the 2008 recipient of the prestigious Karl O. Werwath Engineering Research Award from MSOE. In 2010, he was named a Laser Sintering DINO (Distinguished Innovator Operator Award) from the Additive Manufacturing Users Group and a member of the board of directors for the Wisconsin Manufacturing Extension Partnership (WCMP). Kamara holds the RTAM Master Level Certificate on additive manufacturing from SME and a Bachelor of Science in Mechanical Engineering with Honors from the University of Sierra Leone and a Master of Science in Engineering from MSOE.

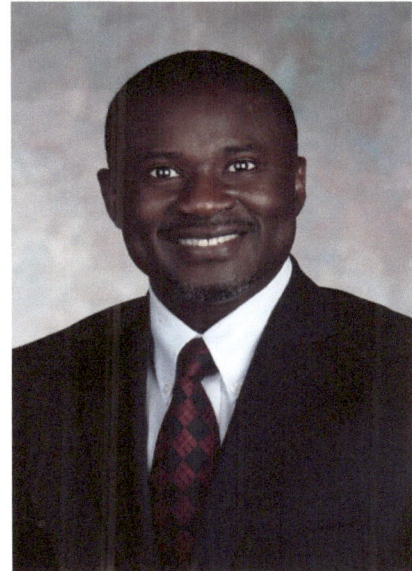

Kamara Sheku
Dean of Applied Technology Center

Other major centers of excellence and cooperative ventures within the ATC include:

- ❖ Professional Education and Research Development (PERD).
- ❖ Rapid Prototyping Center (additive manufacturing).
- ❖ Center for Bio-Molecular Modeling (CBM).
- ❖ Clinical and Translational Science Institute (CTSI).
- ❖ Construction Science and Engineering Center.
- ❖ Engineering Research Center for Compact and Efficient Fluid Power (CCEFP).
- ❖ Mid-West Energy Research Consortium (M-WERC).
- ❖ Nano-Engineering Laboratory.
- ❖ Photonics and Applied Optics Center.
- ❖ Wisconsin Center for Commercialization Resources (WCCR).
- ❖ Wisconsin Space Grant Consortium.

MSOE seminars offer participants the opportunity to explore technological developments and current applications and techniques. The programs are designed to keep practicing engineers abreast of new developments and applications, and also to provide a basic understanding of the technology to new entrants into the field.

MSOE seminars:

- Are based on applied research conducted by scholars.
- Use state-of-the-art laboratories with industrial-grade training equipment.
- Use a hands-on approach to reinforce the concepts presented in class.
- Applications-oriented and often customized to the industry or companies of seminar participants.
- Are offered on the basis of strong long-term partnerships with set objectives and outcomes.

On-site seminars:

MSOE seminars are available for an on-site presentation at your company. The curriculum may be presented in its original format or be modified to meet your specific needs. Confidentiality protected!

MSOE's seminars are unique in the industry because:

- Seminar instructors are experts in their fields, including certified fluid power specialists, Professional Engineers and Ph.D.s.
- Attendees are exposed to the latest fluid power research and industry projects developed at the Fluid Power Institute.
- Professional education seminars use the latest software versions of MATLAB®/Simulink® and Automation Studio in the advanced courses.
- Attendees can network and build professional relationships while benefitting from training, research and industrial projects.

Dr. Medhat Khalil
Director of Professional Education & Research Development (PERD)
www.msoe.edu/seminars
Office: (414) 277-7269
Cell: (414) 940-2232
Fax: (414) 277-7470
khalil@msoe.edu
1025 N. Broadway, Milwaukee, WI, 53202-3109, USA

International **Fluid Power**™ Society

For more than 50 years, MSOE's Fluid Power Institute has been a leader in fluid power research and education. As a part of MSOE's practical, applications-oriented education philosophy, FPI staff and students conduct research and analysis for agriculture, construction, mining off-highway and industrial fluid power applications. Its client list includes global companies such as Caterpillar, CNH, Exxon-Mobil, Husco International, John Deere, Parker Hannifin and Sun Hydraulics.

Premier companies choose the Fluid Power Institute as a partner because of its expertise in evaluating a wide range of hydraulic components and machinery. FPI engineers design and build specialized power supplies and instrumentation systems for pump, motor and fluid efficiency testing. The range of power FPI has available to conduct high-pressure endurance testing of hoses, valves, tubes, plugs and seals is also unique; cylinders as short as a pencil, and as long as a semi-trailer can be evaluated.

The FPI has two facilities that enable it to evaluate a remarkable range of equipment:

On Campus FPI Laboratory
2,400 square foot, Eight test-cells, endurance, fatigue, performance and efficiency tests.

Off-campus FPI Laboratory
12,000 square-feet, high-bay ceiling, drive-in access, reconfigurable workspace and major hydraulic power capabilities.

Timothy Kerrigan
Director of Fluid Power Institute
www.msoe.edu/fpi

FPI's newly established off-campus laboratory is located in the Chase Commerce Center on the south side of Milwaukee. This facility is especially suited for evaluating large components, systems and vehicles. A reconfigurable work space enables FPI engineers to customize power and test conditions using a variety of methods.

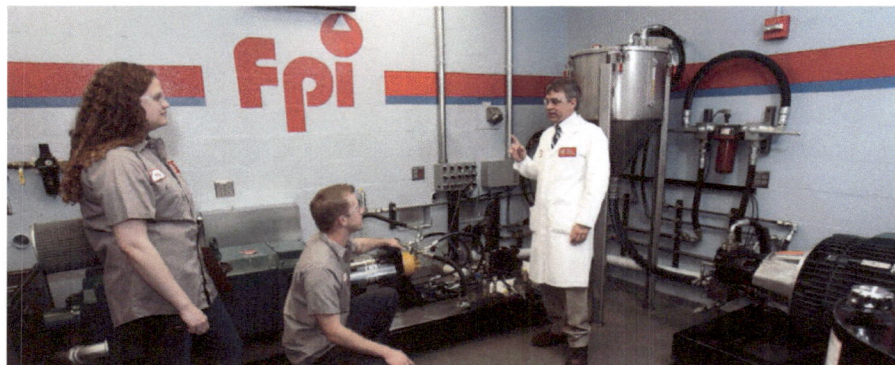

FPI has the ability to design test plans that meet the unique requirements of its clients. Our engineers and students work closely with clients to determine their exact needs. Systems are designed, built and instrumented to test equipment under the appropriate duty cycle. Tests can be conducted according to customer specifications, ASTM, ISO, NFPA or SAE standards.

Engineering Services

The key to developing a reliable, available and maintainable fluid power system is to make it an integral part of the engineering process, and to eliminate failures and failure modes through identification, classification, analysis and removal or mitigation. When developing fluid power systems, it is imperative to select the right activities and to conduct those activities at the right time. The engineering faculty and staff at FPI are experts in fluid power application from a simple design to an efficient and reliable hydraulic or pneumatic system.

Tribology Services

The FPI has been a leader in contamination analysis and filtration technology for decades. In the 1980's, FPI pioneered the use of automatic particle counters in hydraulic fluid analysis. In the 1990's, FPI pioneered the development of surgically clean fluids for initial-fill applications. In the 2000's FPI was the very first to use Atomic Force Microscopy in wear particle analysis. FPI's role as a practitioner and educator in these areas has truly advanced the fluid power industry. Our current research thrust incorporates the study and formulation of energy-efficient hydraulic fluids-an endeavor funded by a grant from the National Science Foundation and industry partners.

Many of the world's largest equipment manufacturers use FPI to test new hoses, tubes, cylinders, coolers, reservoirs, pumps, bearings and valve assemblies to determine the type and size of manufacturing contamination, left in the component as received by the customer. Through the use of advanced diagnostic methods such as ferrography, atomic force microscopy, stereomicroscopy and laser particle imaging, early detection and root-cause analysis are possible.

Dr. Medhat Khalil
Director of Professional Education
and Research Development
1025 North Broadway
Milwaukee, WI 53202-3109
(414) 277-7269
khalil@msoe.edu
www.msoe.edu/seminars

Medhat Khalil, Ph.D., Director of Professional Education & Research Development at the Applied Technology Center, Milwaukee School of Engineering, Milwaukee, WI, USA. Medhat got his bachelor and master's degree in mechanical engineering from Cairo, Egypt. Medhat has been granted his Ph.D. in Mechanical Engineering and Post-Doctoral Industrial Research Fellowship from Concordia University in Montreal, Quebec, Canada. Medhat, so far, published a couple of textbooks. Medhat participated in many technical conferences and published a good number of reviewed technical papers and he is in the process of registering number of patents. Medhat has been certified by the International Fluid Power Society (IFPS) as: Certified Fluid Power Hydraulic Specialist (CFPHS) and Certified Fluid Power Accredited Instructor (CFPAI). Medhat is a member of many grand institutions such as Center for Compact and Efficient Fluid Power Engineering Research Center (CCEFP), listed Fluid Power Consultant by the National Fluid Power Association (NFPA), listed professional instructor by the American Society of Mechanical Engineers (ASME), and listed professional instructor by the National American Die Casting Association (NADCA). Medhat has been assigned as the chair of the education committee for the International Fluid Power Exposition (IFPE2017 and 2020). Medhat developed and taught various courses for industry professionals. Medhat has a balanced academic and industrial experience. Medhat has a deep working experience in the field of Mechanical Engineering; more specifically in fluid power and motion control. Medhat had worked for several world-wide recognized industrial organizations such as Rexroth in Germany and CAE in Canada. Medhat had designed several hydraulic systems and developed several analytical and educational software. Medhat also has vast experience in modeling and simulation of dynamic systems using Matlab-Simulink.

MSOE UNIVERSITY

Thomas S. Wanke, CFPE Director Fluid Power Industrial Consortium and Industry Relations

Tom is the Director of the MSOE's new Fluid Power Industrial Consortium and Industry Relations. Previously he was Director of MSOE's Fluid Power Institute for 37 years. He is also an adjunct assistant professor in MSOE's Mechanical Engineering Dept. Tom is an IFPS Certified Fluid Power Engineer and a Certified Fluid Power Specialist. He has over 48 years of experience in the fluid power industry, 45 of which have been at MSOE. Tom has a bachelor's degree in mechanical engineering technology with a fluid power focus and a master's degree in engineering with a fluid power specialty option both from MSOE. He has worked on numerous projects in the following areas: component and system design; development and evaluation; field troubleshooting and failure analysis; fluids, filtration and contamination control.

Tom is a member IFPS and is past Co-Chairman of ISO TC131 SC8; NFPA T2.12; and NFPAT2.24 Standards Committees. He is also past chairman of the NFPA Technical Board, IFPE 2011 and 2014 and Educational Programs. Tom has written and presented various technical papers at conferences; taught numerous fluid power classes, seminars and short courses.

Contact:

Thomas S. Wanke, CFPE
Director MSOE Fluid Power Industrial Consortium and Industry Relations
1025 North Broadway
Milwaukee, WI 53202-3109
(414) 277-7191
wanke@msoe.edu
www.msoe.edu

Dr. Daniel Williams

is an associate professor in MSOE's Mechanical Engineering department. He earned his bachelor's degree in mechanical engineering from the University of Wisconsin-Platteville and his master's degree and Ph.D. in mechanical engineering from the University of Wisconsin-Madison. Williams has more than 20 years of industry engineering experience. He worked for two years as a design engineer at Snap-On Tools Corporation in Kenosha, Wis. Following graduate studies, Williams worked for 18 years in John Deere's Construction & Forestry Division in Dubuque, Iowa, where he specialized in machine systems simulation—hydraulics, drive train, rigid body dynamics and controls—and control design. Dan has also been a member of the full-time faculty at Loras College in Dubuque, where he taught courses in the electromechanical engineering program for five years.

Paul Michael, C.L.S., is a research chemist in MSOE's Fluid Power Institute. He earned his B.S. in chemistry at the University of Wisconsin, Milwaukee and graduated with distinction from Keller Graduate School of Management. He has more than 30 years of experience in the formulation and testing of hydraulic fluids and lubricants. Paul is an STLE certified Lubrication Specialist and chairs the NFPA Fluids Committee. In addition to his research in contamination analysis, he is currently investigating energy efficient hydraulic fluids in the NSF funded multi-university Center for Compact and Efficient Fluid Power. Michael was a recipient of the Otto J. Maha Pioneers in Fluid Power Award in 2012.

Milwaukee: A fluid power industrial hub

The following fluid power related component manufacturers, machine builder, service providers, associations and organization are samples of Milwaukee-based or at least have a subsidiary in the city of Milwaukee:

- ❖ Actuant.
- ❖ Caterpillar Mining.
- ❖ CASE.
- ❖ Eaton R&D.
- ❖ Fluid System Components.
- ❖ Grimstad.
- ❖ GS- Hydraulics.
- ❖ Husco International.
- ❖ Milwaukee Cylinders.
- ❖ Milwaukee School of Engineering.
- ❖ Motion Industries.
- ❖ Milwaukee Hydraulics.
- ❖ National Fluid Power Association (NFPA).
- ❖ Norman Equipment.
- ❖ Oilgear.
- ❖ Poclain Hydraulics.
- ❖ P & H Mining.
- ❖ PUTZMEISTER America.
- ❖ Racine Federated.

MSOE
UNIVERSITY

Certification and Continuing Education Credit Units

The Professional Education Department authorizes Continuing Education Credit Unis (CEUs) for seminar participants. For every 1 contact hour, 0.1 CEU is granted. For example, 10 Hours seminar participant deserve 1 CEU.

For an institution to be eligible to grant CEU, it must meet certain criteria:
❖ It should be a recognized institution.
❖ Courses must be with learning objectives.
❖ Presenter must be qualified instructor.
❖ Records of participants must be kept and maintained.
❖ Participants must be certified either by a certification exam or hands-on practice.

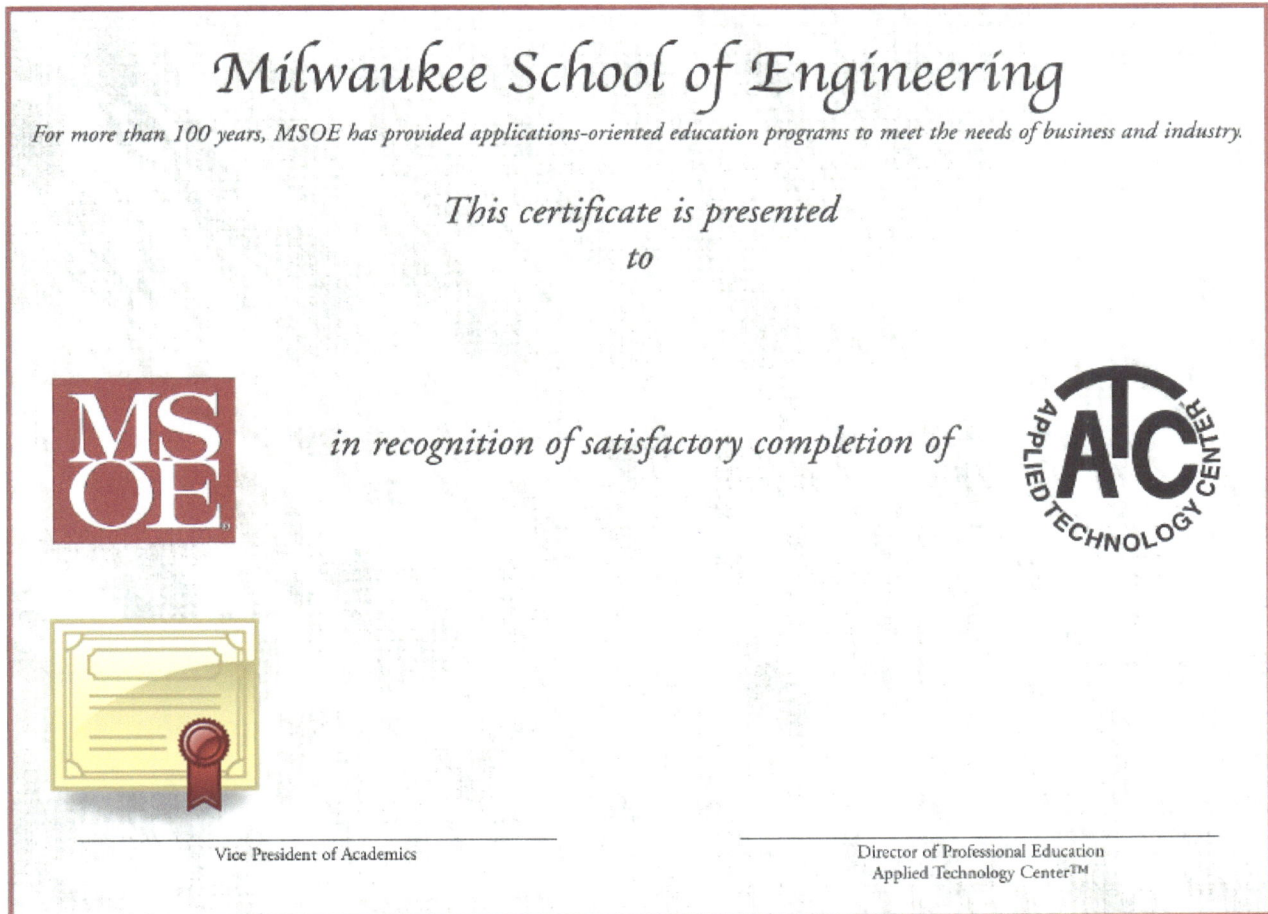

Milwaukee School of Engineering

For more than 100 years, MSOE has provided applications-oriented education programs to meet the needs of business and industry.

This certificate is presented
to

in recognition of satisfactory completion of

_____
Vice President of Academics

_____
Director of Professional Education
Applied Technology Center™

Targeted Clients

We provide quality training for professionals from
Industrial and Mobile Applications.

- ❖ Steelworks and Metal Forming.
- ❖ Molding and Die Casting Industries.
- ❖ Forging and Extrusion Works.
- ❖ Machine Tools.
- ❖ General Manufacturing.
- ❖ Industrial Automation.
- ❖ Process Engineering.
- ❖ Wood, Paper and Glass Industries.
- ❖ Pharmaceutical Industries.
- ❖ Chemical and Petrochemicals.
- ❖ Food Industries.
- ❖ Power Plants.
- ❖ Renewable Energy.
- ❖ Material Handling.

- ❖ Earth Moving Machines.
- ❖ Construction Machines.
- ❖ Lifting Equipment.
- ❖ Agricultural Machines.
- ❖ Oil & Gas Industries.
- ❖ Offshore Equipment.
- ❖ Mining Equipment.
- ❖ Marines & Shipbuilding.
- ❖ Defense Systems.
- ❖ Aerospace Industries and
- ❖ Airport Service Machines.
- ❖ Rail-way Vehicles.
- ❖ City service Vehicles.
- ❖ Automotive engineering.

MSOE UNIVERSITY

Training Equipment

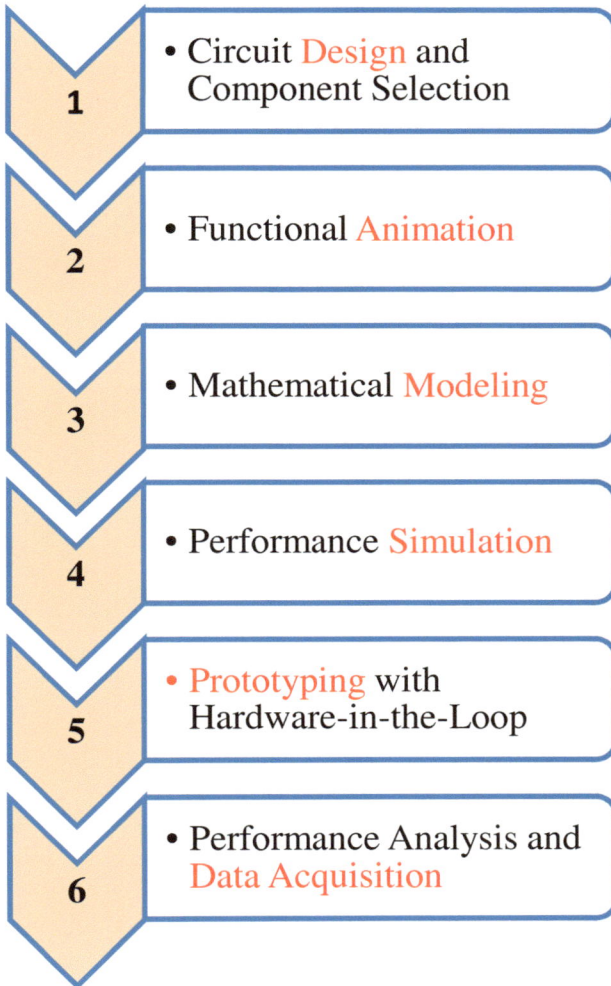

1
• Circuit Design and Component Selection

2
• Functional Animation

3
• Mathematical Modeling

4
• Performance Simulation

5
• Prototyping with Hardware-in-the-Loop

6
• Performance Analysis and Data Acquisition

The state-of-the-art Universal Fluid Power Trainer (UFPT) has been designed by Dr. Khalil. Four fully functional units have been added to the department of Professional Education to be used by seminar participants to practice designing, animating, simulating and building hydraulic circuits. The machines are universal, transportable and compact so that it can be shipped to the customer's site.

To learn more about the Universal Fluid Power Trainers:
http://www.msoe.edu/seminars

Animate it

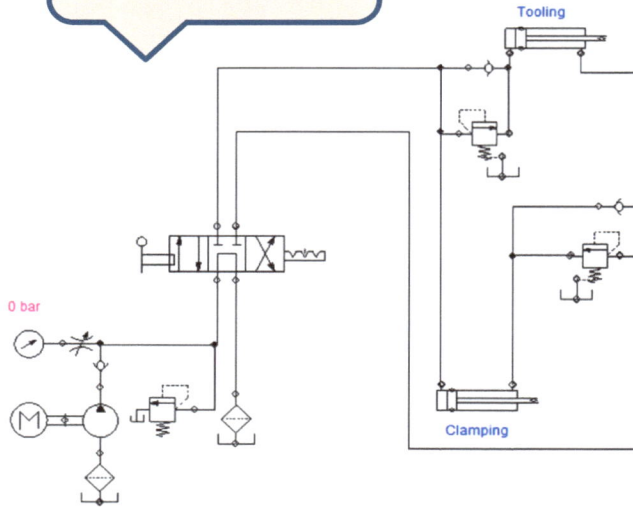

Practice it

Simulate it

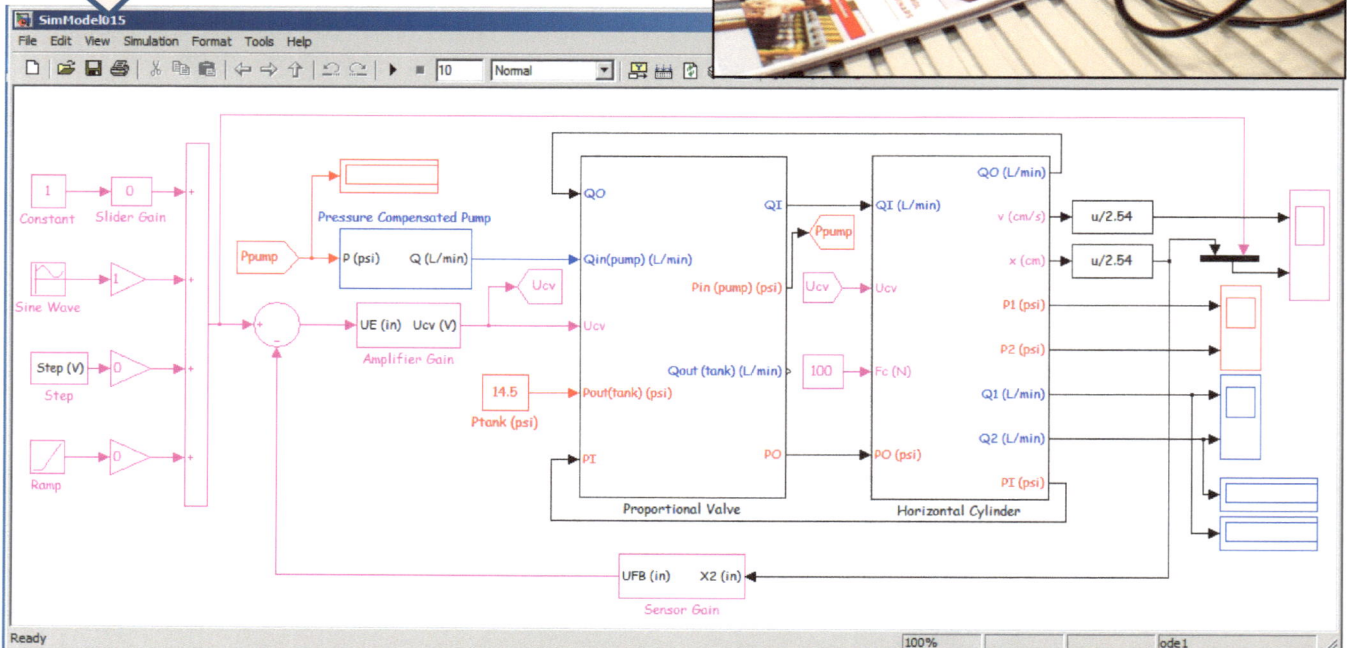

Photo Gallery

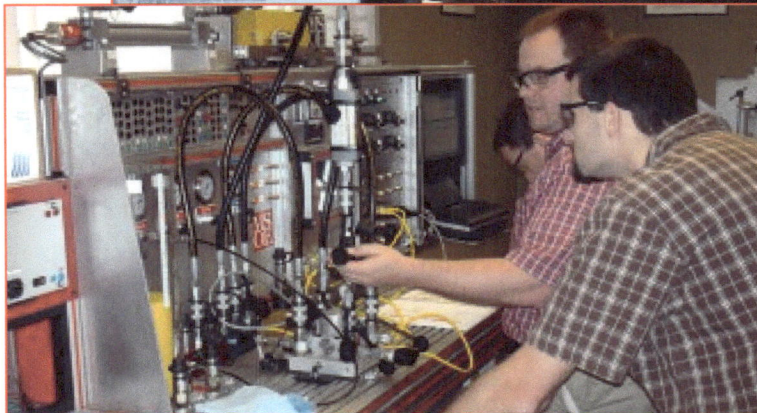

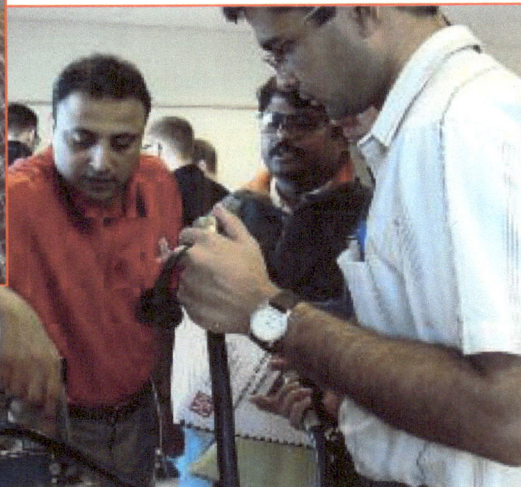

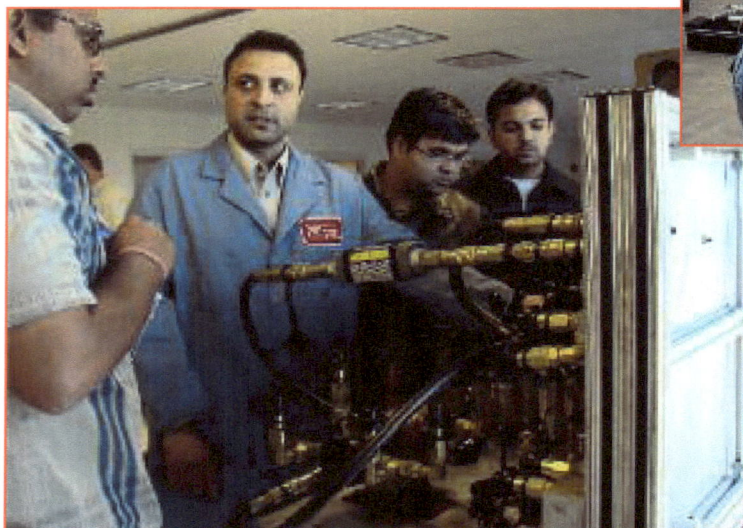

Logistics

Air Travel to Milwaukee:
Book your flight to Mitchell International Airport (Airport Code: MKE), it is a 15-minute taxi ride to downtown Milwaukee.

Pleases Review your Confirmation Letter
Public Classes are held in one of the following two locations

Location 1 – MSOE Campus

- 429 E. State Street, Milwaukee, WI 53202. MAP
- Training room # S100 on the first floor in the Science Building.
- Contact: Dr. Medhat Khalil Tel: 1-414-940-2232.

Location 2 - GS Global Resources

- 926 Perkins Drive, Mukwonago, WI 53149. MAP
- Training room # 5050 on the second floor.
- Contact: Jeanette Cutberth Tel: 1-262-378-5225.

Dressing:
Dress casual and comfortable. Look up Milwaukee weather forecast to plan your trip. www.weather.com, zip code: Mukwonago, WI 53149.

Parking:
Seminar participants will be given free parking spots.

MSOE UNIVERSITY

Where to Stay:

The following are the hotel recommendation around the seminar location:

Location 1 – MSOE Campus

Hyatt Regency Milwaukee
1234-276 (414)
1234-233 (800)
333 W. Kilbourn Ave.
$119 plus tax per night
www.hyatt.com

The Astor Hotel
4220-271 (414)
924 E. Juneau Ave.
$69 per week night plus tax
www.thehotelastor.com

Location 2 – GS Global Resources

La QUINTA Inns & Suites
15300 West Rock Ridge Road
New Berlin, WI 53151
1-262-717-0900
Rates are usually around $89.00. GS Global Resources has a Corporate Rate of 20% discount. When you call the above number you will get an auto attendant, press 6 for the front desk to ask for the GS Global Resources rate.

Holiday Inn Express & Suites
15451 W. Beloit Road
New Berlin, WI 53151
1-800-392-1019
Rates are usually around $133.00. GS Global Resources has a Corporate Rate of $84. When you call to make your reservation, please mention this special rate.

Quality Inn & Suites
2929 O'Leary Lane
East Troy, WI 53120
1-262-642-2100
Ask about GS Global Resources Corporate Rate.

Eagle Centre House Bed and Breakfast
W370 S9590, Hwy 67
Eagle, WI 53119
262-363-4700

Seminar Tree

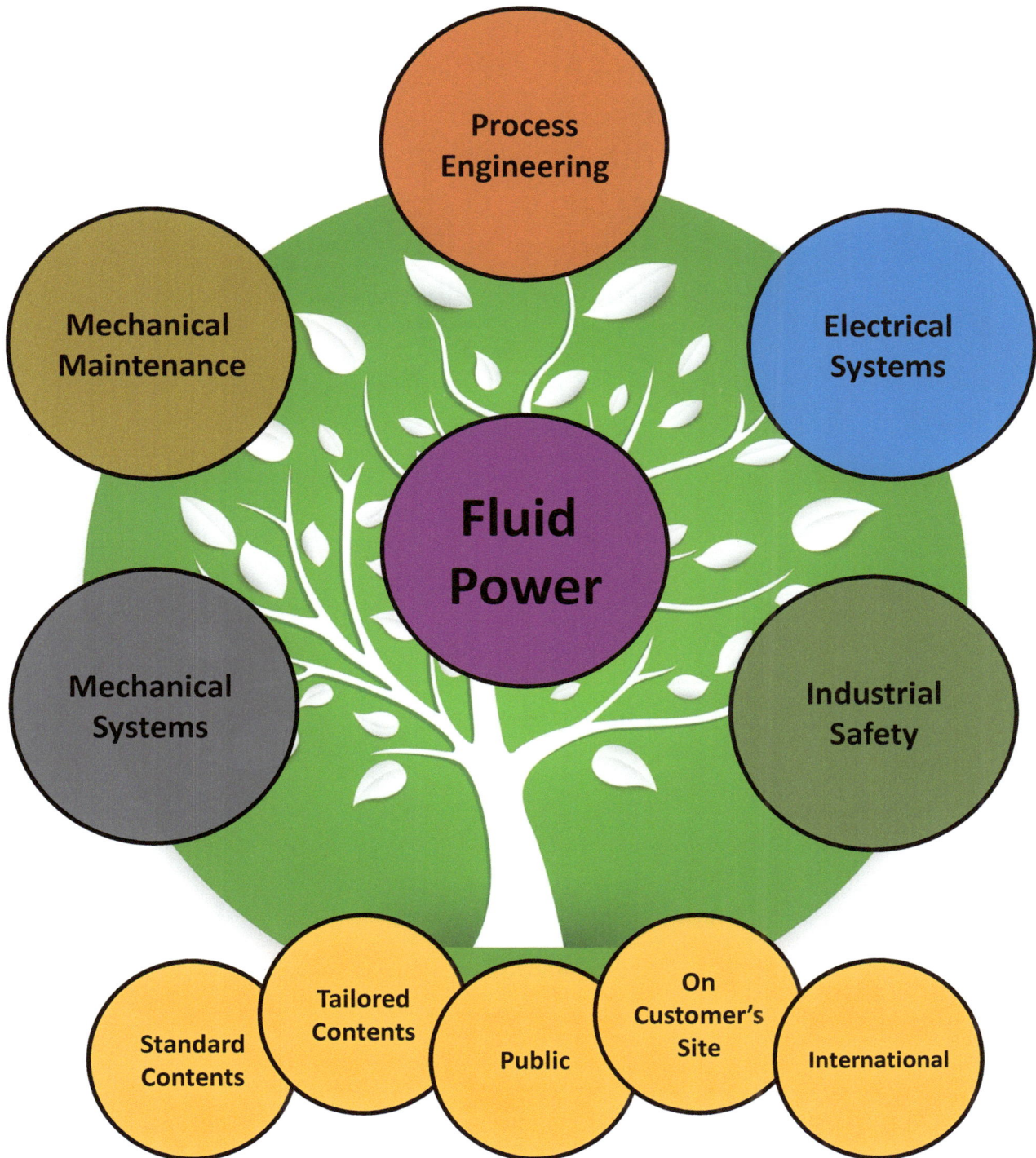

- Process Engineering
- Mechanical Maintenance
- Electrical Systems
- Fluid Power
- Mechanical Systems
- Industrial Safety
- Standard Contents
- Tailored Contents
- Public
- On Customer's Site
- International

Registration

Fax to: +1-414-277-7470

Mail to (Att. Dr. Medhat Khalil)
Applied Technology Center
Milwaukee School of Engineering.
1025 N. Broadway, Milwaukee, WI, 53202-3129

Call: +1-414-277-7195 **OR:** +1-414-277-7269

www.msoe.edu/seminars

khalil@msoe.edu

Cancellation Policy:
- MSOE reserves the right to cancel a seminar if minimum enrollment is not met.
- Cancellation from the client side subject to the following conditions:
  - Three weeks before the seminar date are subject to a $200 cancellation fee with a refund of the remainder.
  - Cancellations two weeks before the seminar date are subject to a $400 cancellation fee with a refund of the remainder.
  - Cancellations one week before the seminar date are subject to a $600 cancellation fee and the remaining funds will be used as a credit towards any future seminar, subject to availability.

Seminar Registration Form
Please enroll the individual (s) listed below in:

Seminar #	Seminar Name	Seminar Date

Name: ---

Title: ---

Tel: ---

Email: ---

Company: ---

Address: --

Payment Method:
   Company Purchase Order: PO# -------------
   Charge Seminar Fee (s) to   Master   VISA   Discover   A. Express

Account Number -- Expiration Date: --------------------

Name on the Card   Signature

MSOE-PERD   www.msoe.edu/seminars   khalil@msoe.edu   Cell: +1-414-940-2232

Department of Professional Education

Seminar Registration Form
Please enroll the individual (s) listed below in:

Seminar #	Seminar Name	Seminar Date

Name: --

Title: ---

Tel: --

Email: ---

Company: ---

Address: --

Payment Method:
   Company Purchase Order: PO# -------------
   Charge Seminar Fee (s) to   Master   VISA   Discover   A. Express

Account Number -- Expiration Date: -------------------

Name on the Card   Signature

MSOE-PERD   www.msoe.edu/seminars   khalil@msoe.edu   Cell: +1-414-940-2232

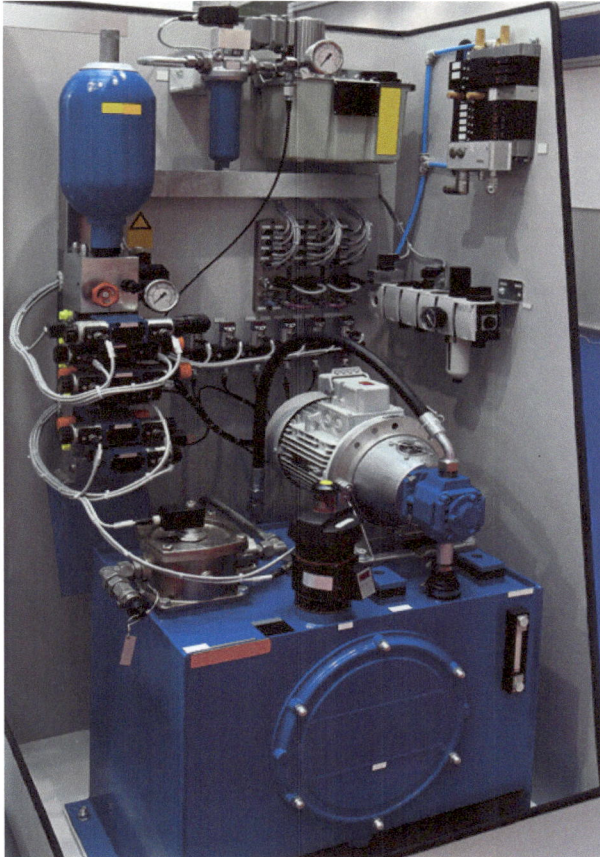

We serve both industrial and mobile applications!

We are the sole provider of fluid power training to largest construction companies nationwide

We are member of:

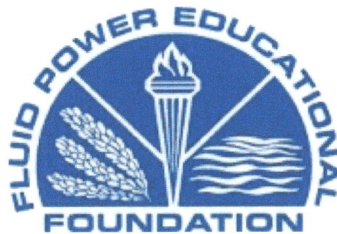

FLUID POWER EDUCATIONAL FOUNDATION

International **Fluid Power** Society

NATIONAL FLUID POWER ASSOCIATION · NFPA ·

CENTER FOR COMPACT AND EFFICIENT FLUID POWER
NSF A National Science Foundation Engineering Research Center

MSOE-PERD   www.msoe.edu/seminars   khalil@msoe.edu   Cell: +1-414-940-2232

Designation Table

Condition	Des.	Clarification
Exam:	✓	Course contains certification exam to get certified
	✗	No certification exam.
Hands-On:	✓	Course contains hands-on labs.
	✗	Course conducted on theoretical base.
Scheduled:	✓	Course scheduled and registration is opened for public
	✗	Course is offered upon request at the customer-site or for public when the minimum enrollment number is reached.
	UD	Course is under development.

Customize Your Own Industrial Training. Courses can be mobilized to your facility.
For dates and locations of courses scheduled in current calendar year visit: www.msoe.edu/semiars

Fluid Power Training

Course #	Course Title	CEU	Hr	Days	Hands-On	Exam	Scheduled	$/Person
MSOE-H00	Hydraulic Specialist Certification Review Session	1.8	18	3	✗	✓	✓	$600
MSOE-H01	Introduction to Hydraulic Systems for Industry Professionals	2.7	27	5	✓	✗	✓	$2,160
MSOE-H02	Electro-Hydraulic Components and System	2.7	27	5	✓	✗	✓	$2,160
MSOE-H03	Hydraulic Fluids and Contamination Control	1.2	12	2	✗	✗	✓	$600
MSOE-H04	Hydraulic Fluids Conditioning	1.2	12	2	✗	✗	UD	$600
MSOE-H05	Safety and Maintenance	1.2	12	2	✗	✗	UD	$600
MSOE-H06	Troubleshooting and Failure Analysis	1.2	12	2	✗	✗	UD	$600
MSOE-H07	Modelling and Simulation for Application Engineers	2.7	27	5	✓	✗	✓	$2,160
MSOE-H08	Design Strategies of Hydraulic Systems	2.7	27	5	✓	✗	UD	$2,160
MSOE-H09	Design Strategies for Electro-Hydraulic Systems	2.7	27	5	✓	✗	UD	$2,160
MSOE-HF01	Fundamentals of Hydraulic Systems	0.5	5	1	✗	✗	✗	$300
MSOE-HF02	Fundamentals of Electro-Hydraulic Systems	0.5	5	1	✗	✗	✗	$300
MSOE-HF03	Fundamentals of Service and Operation of Hyd. Sys.	0.5	5	1	✗	✗	✗	$300
WMK01	Overview of Fluid Power Systems	0.6	6	1	✗	✗	✗	$300
WMK02	Fluid Power Applications	1.2	12	2	✗	✗	✗	$600
WMK03	Hydraulic Motors Construction and Application	1.2	12	2	✗	✗	✗	$600
MSOE-P00	Pneumatic Specialist Certification Review Session	1.8	18	3	✗	✓	✗	$600
MSOE-P01	Introduction to Pneumatic Systems for Industry Professionals	1.2	12	2	✓	✗	UD	$600
MSOE-P02	Electro-Pneumatic Components and Systems	1.2	12	2	✓	✗	UD	$600

Fluid Power Training

Hydraulic Course Map

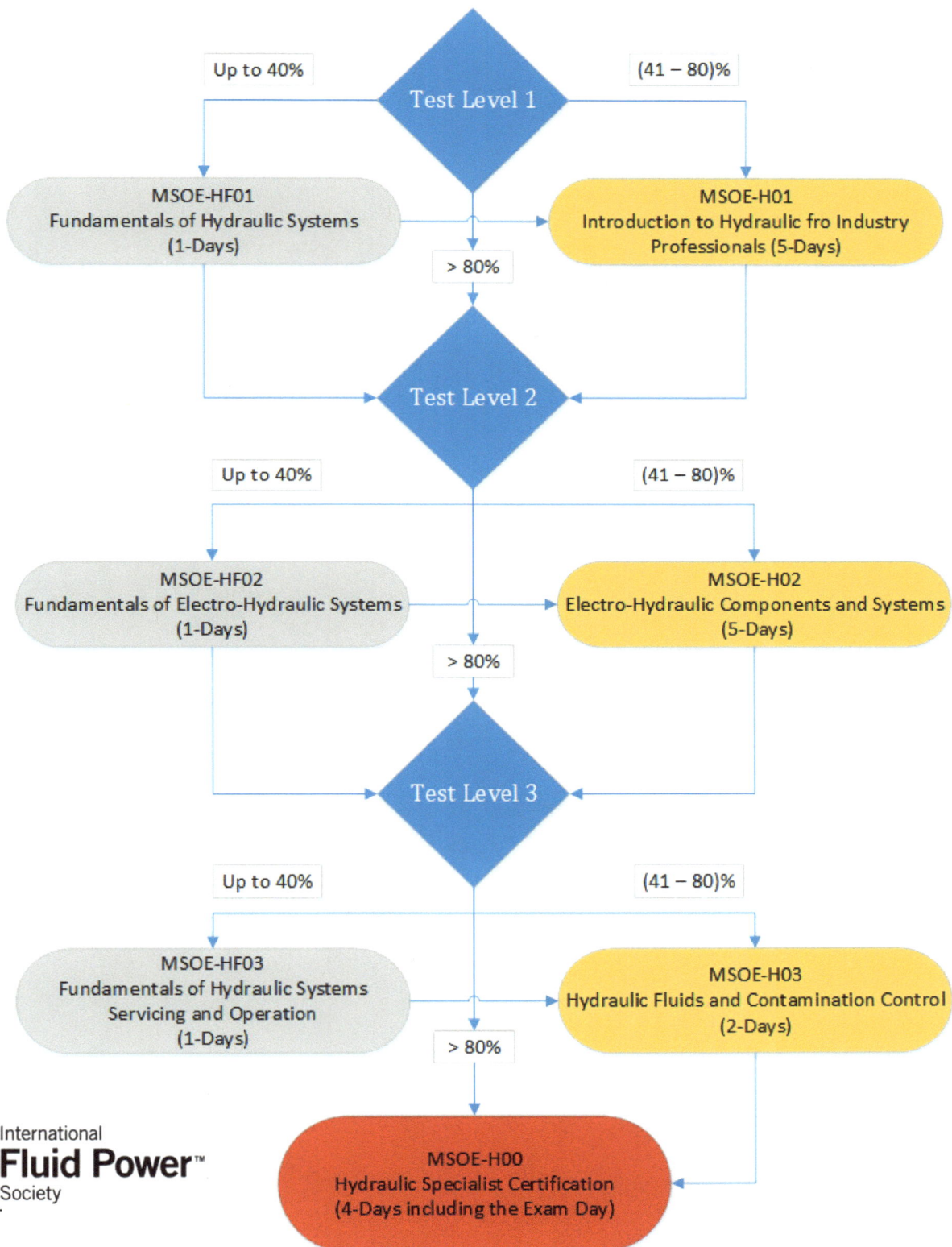

Test Level 1

Up to 40% → MSOE-HF01 Fundamentals of Hydraulic Systems (1-Days)

(41 – 80)% → MSOE-H01 Introduction to Hydraulic fro Industry Professionals (5-Days)

> 80%

Test Level 2

Up to 40% → MSOE-HF02 Fundamentals of Electro-Hydraulic Systems (1-Days)

(41 – 80)% → MSOE-H02 Electro-Hydraulic Components and Systems (5-Days)

> 80%

Test Level 3

Up to 40% → MSOE-HF03 Fundamentals of Hydraulic Systems Servicing and Operation (1-Days)

(41 – 80)% → MSOE-H03 Hydraulic Fluids and Contamination Control (2-Days)

> 80%

MSOE-H00 Hydraulic Specialist Certification (4-Days including the Exam Day)

International **Fluid Power** Society

Course Map for Pneumatic

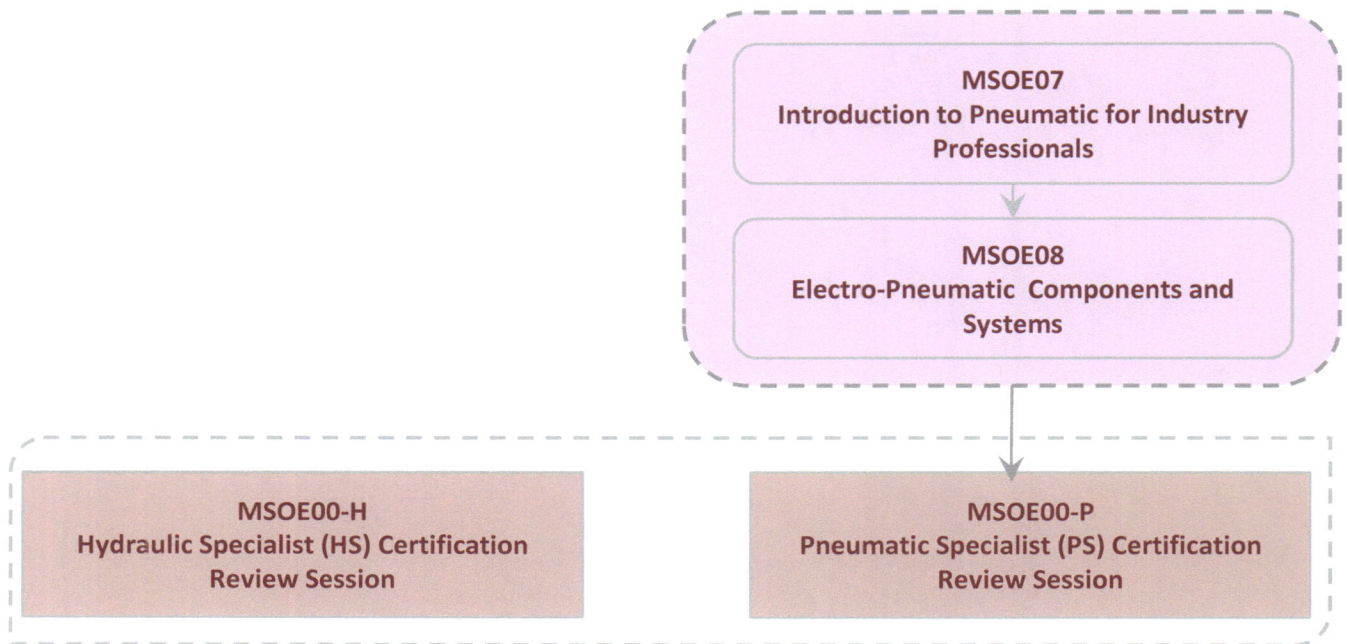

> **MSOE07**
> **Introduction to Pneumatic for Industry Professionals**
>
> ↓
>
> **MSOE08**
> **Electro-Pneumatic Components and Systems**

MSOE00-H
Hydraulic Specialist (HS) Certification Review Session

MSOE00-P
Pneumatic Specialist (PS) Certification Review Session

Once an individual completes HS and PS certification levels, they are considered a Fluid Power Specialist; no additional written test is require

International
Fluid Power™
Society

Hydraulic Specialist Certification Review Session

Fluid Power Training

Course #	Course Title	CEU	Hr	Days	Hands-On	Exam	Scheduled	$/Person
MSOE-H00	Hydraulic Specialist Certification Review Session (IFPS Certification)	1.8	18	3	✗	✓	✓	$600

Course Description:

This 18-hours 3-days review session is conducted at MSOE followed by the certification exam on the fourth day. The objective of the course is to walk the participants through the study manual provided by IFPS in order to maximize their chance to pass the certification exam.

What is the IFPS?

The International Fluid Power Society is the only organization that offers comprehensive technical certification for all professionals in the field of fluid power and motion control industry.

What is the Process of Certification?

After 3-days review session provided by MSOE, participants will take the certification exam on fourth day. Exam will be provided and proctored by IFPS. The test is 3-hours, 50-questions, and multiple-choice type of test. You need to get 35 correct answers out of 50 questions. If you fail, you can re-schedule taking the exam at a later time. If you pass, you will be issued a "Hydraulic Specialist" certificate. The certificate is good for five years, after five years you do not need to retake the exam, you need only to report to IFPS indicating that you are still working in the field.

Why Get Certified?

- The "Hydraulic Specialist" certification is an internationally recognized certification.
- The certificate is portable - it goes with the individual wherever they work.
- Certifications help an individual to advance his career and introduce himself to the global job market.
- Certification sets an individual apart as a leader in their chosen field of work.
- Certification will help a vendor provide over-the-top quality and acquire ISO certification easily.
- Certified personnel help make the work environment safe and improve the safety, reliability and efficiency of a machine operation.

MSOE
UNIVERSITY

Course Agenda

Course Agenda: AM Session (9-Noon)   Lunch Hour (Noon - 1 pm)   PM Session (1 - 4)

	Day 1	Hr
AM	Registration and Orientation Session	0.5
	Job Responsibility 1: Understand the Function of Hydraulic Components in Circuits	2.5
PM	**Job Responsibility 1:** Continue	1.5
	Job Responsibility 2: Analyze Loads and Motion	1.5
	Day 2	Hr
AM	**Job Responsibility 3: Select Components for Hydraulic Systems**	3.0
PM	**Job Responsibility 4:** Analyze and Troubleshoot Hydraulic Systems	1.5
	Job Responsibility 5: Electrohydraulic Control Systems	1.5
	Day 3	
AM	**Pretest 1 and 2**	3.0
PM	**Pretest 3 and 4**	3.0
	Total	18
	Day 4	
AM	**Certification Exam**	3

MSOE-PERD   www.msoe.edu/seminars   khalil@msoe.edu   Cell: +1-414-940-2232

Introduction to Hydraulic Systems for Application Engineers

Fluid Power Training

Course #	Course Title	CEU	Hr	Days	Hands-On	Exam	Scheduled	$/Person
MSOE-H01	Introduction to Hydraulic Systems for Application Engineers	2.7	27	5	✓	✗	✓	$2160

Course Description:

This 27-hour 5-day seminar is designed to acquaint individuals with the fluid power field and provide a practical working knowledge of this important and growing industry. This program features laboratory sessions where participants will gain practical experience working with actual fluid power components and systems. Specifically, laboratory sessions will treat the disassembly, inspection and assembly of individual components, as well as system design examples. This class explores not only how hydraulic components work, but why it works this way.

Seminar attendees will receive:
- The shown textbook.
- A workbook.
- Download the "Hydraulic Component Sizing Calculator".
- Download the animated circuits.

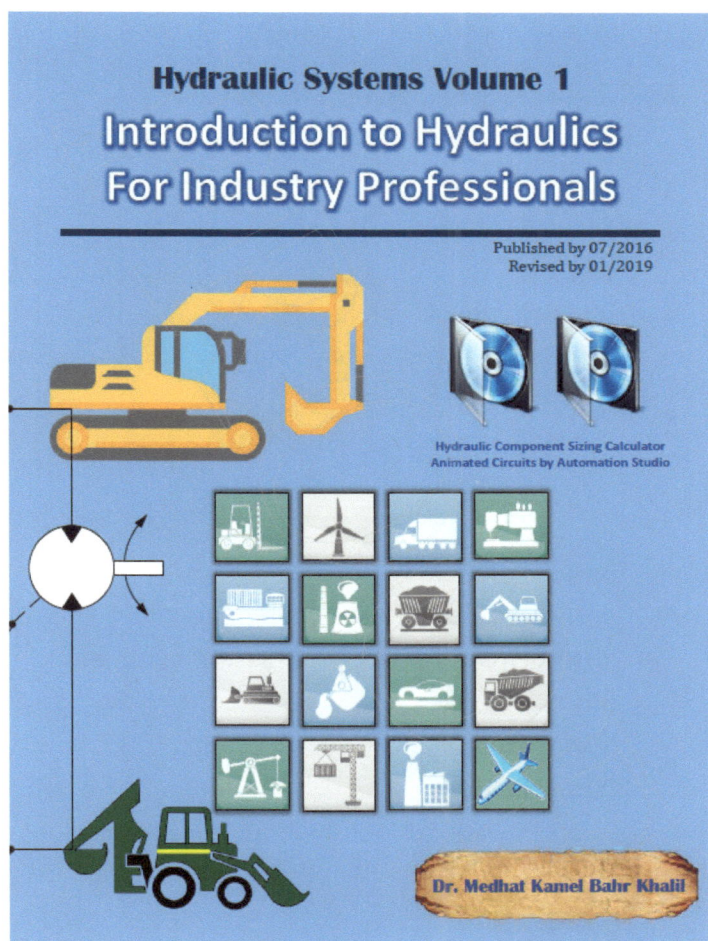

Hydraulic Systems Volume 1
Introduction to Hydraulics For Industry Professionals

Published by 07/2016
Revised by 01/2019

Hydraulic Component Sizing Calculator
Animated Circuits by Automation Studio

Dr. Medhat Kamel Bahr Khalil

ISBN-13: 978-0692622360

Fluid Power Training

Course Agenda

Course Agenda: AM Session (9-Noon)   Lunch Hour (Noon - 1 pm)   PM Session (1 - 4)

	Day 1	Hr
	Registration and Orientation Session	0.5
AM	**CH01:** Hydraulic Systems Overview	1.5
	CH02: Basic Concepts Review	1
	CH02: Contd.	1.5
PM	**Lab 1:** Energy Losses in Hydraulic Conductors	1
	CH03: Hydraulic Component Sizing Calculations	0.5
	Day 2	**Hr**
AM	**CH04:** Hydraulic Pumps and Motors	3
	Lab 2: Power Distribution in a Hydraulic System	1
PM	**CH04:** Contd.	1.5
	Inspect Pumps and Motors	0.5
	Day 3	**Hr**
AM	**CH05:** Hydraulic Valves Overview	3
	Lab 3: Valve Coefficient Development.	0.5
PM	**CH05:** Contd.	0.5
	CH06: Hydraulic Linear and Rotary Actuators.	1.5
	Lab 4: Motion Control of Hydraulic Cylinder.	0.5
	Day 4	**Hr**
	CH07: Hydraulic Accumulators.	0.5
AM	Inspect Valves, Actuators and Accessories.	0.5
	CH08: Hydraulic Circuits for Basic Applications.	2
	CH08: Contd.	0.5
PM	**Lab 5:** Control of Overrunning Loads.	1
	CH08: Contd.	0.5
	Lab 6: Speed Control of Hydraulic Actuator.	1
	Day 5	**Hr**
	CH08: Contd.	1
	Lab 7: Boosting Speed of Hydraulic Cylinder.	1
AM	**CH08:** Contd.	0.5
	Lab 8: Sequence Operation of Hydraulic Cylinder.	0.5
	Machine shutdown Procedure.	
	Total	**27**

MSOE-PERD   www.msoe.edu/seminars   khalil@msoe.edu   Cell: +1-414-940-2232

Electro-Hydraulic Components and Systems

Fluid Power Training

Course #	Course Title	CEU	Hr	Days	Hands-On	Exam	Scheduled	$/Person
MSOE-H02	Electro-Hydraulic Components and System	2.7	27	5	✓	✗	✓	$2,160

Course Description:

This 27-hour 5-day seminar is designed to cover the knowledge of electrohydraulic components and systems. The state-of-the-art Universal Fluid Power Trainers are used to demonstrate the theory presented. The introductory part of the course covers the applications of electro-hydraulic systems and the benefits of converting the classical hydro-mechanical solutions into electro-hydraulic solutions. The core part of the course covers the knowledge of electrohydraulic valves including solenoid operated valves, proportional valves and servo valves. The course also covers the basic functions that are built on the electrical control units and drivers for such valves, e.g., gain adjustor, overload protection, null adjustment, ramp generator, dead-band eliminator, dither, and pulse width modulation. The course also discusses system design considerations and the technicalities of in-field tuning of open-loop and closed-loop electro-hydraulic systems.

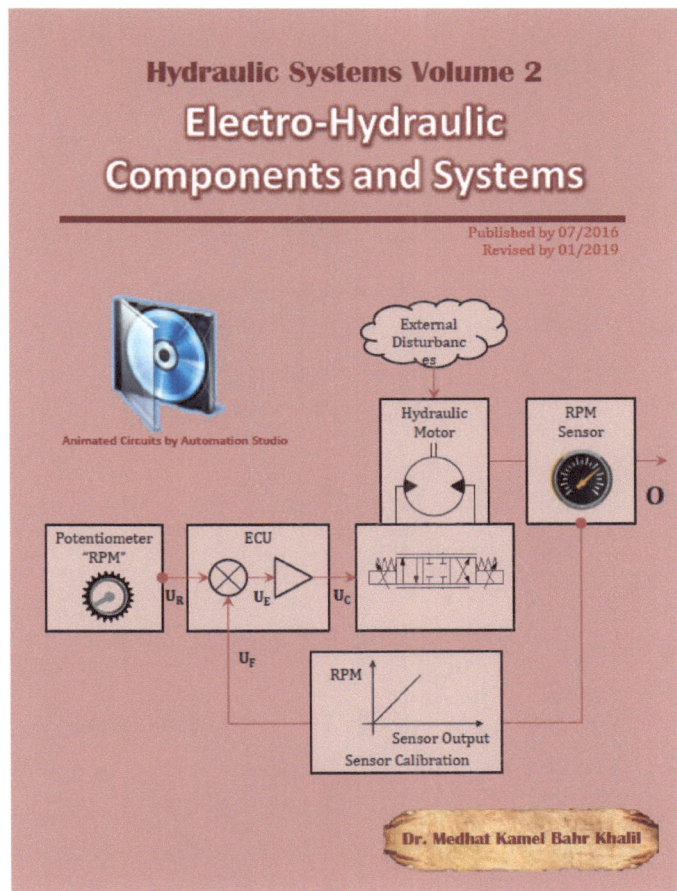

Hydraulic Systems Volume 2
Electro-Hydraulic Components and Systems

Published by 07/2016
Revised by 01/2019

Animated Circuits by Automation Studio

Dr. Medhat Kamel Bahr Khalil

ISBN-13: 978-0997763423

Seminar attendees will receive:
- The shown textbook.
- A workbook.
- Download the animated circuits.

Fluid Power Training

Course Agenda

Course Agenda: AM Session (9-Noon) Lunch Hour (Noon - 1 pm) PM Session (1 - 4)

	Day 1	Hr
AM	Registration and Orientation Session	0.25
	CH01: Hydro-Mechanical vs. Electro-Hydraulic Solutions	2.25
	CH02: Electro-hydraulic System Application	0.5
PM	**CH03:** Switching (ON/OFF) Valves-Construction and Operation	2
	Lab Manual - UFPT	0.5
	Lab20: Cylinder extension upon pressing a push-button	0.5
	Day 2	**Hr**
AM	**CH04:** Switching (ON/OFF) Valves-Circuits for Basic Functions	1.5
	Lab21: Signal storage by electrical self-locking	0.25
	Lab22: Electrical locking by means of contactors contact	0.5
	Lab23: Position-dependent cylinder deceleration	0.5
	Lab24: Pressure-Dependent cylinder reversal	0.25
PM	**Lab25:** Event-Dependent warning circuit	0.25
	CH04: Contd. (Practice building circuits without instructor's orientation).	1.25
	CH05: Proportional Valves	1.5
	Day 3	**Hr**
AM	**CH05:** Contd.	2
	Lab26: Cylinder Motion Control Performance using Switching Valve vs. Prop.Valve	0.5
	CH06: Servo Valves	0.5
PM	**CH06:** Contd.	2.5
	Lab27: Cylinder Motion Control Performance using Servo Valve vs. Proportional Valve	0.5
	Day 4	**Hr**
AM	**CH07:** Electro-hydraulic System Design Considerations	1
	Lab28: Digital Control of EH Variable Displacement Pumps	0.5
	CH08: Control Electronics for Electro-Hydraulic Systems	1.5
PM	**CH08:** Contd.	1
	Lab29: Digital Control of E.Hydraulic Cylinder Position + Machine Shutdown Procedure	1
	CH09: Valve Selection for an Electro-Hydraulic Controlled Actuator	1
	Day 5	**Hr**
AM	**CH09:** Contd.	2
	CH10: Electro-hydraulic Valves Commissioning and Maintenance	1
	Total	27

MSOE-PERD   www.msoe.edu/seminars   khalil@msoe.edu   Cell: +1-414-940-2232

Hydraulic Fluids Conditioning and Contamination Control

Fluid Power Training

Course #	Course Title	CEU	Hr	Days	Hands-On	Exam	Scheduled	$/Person
MSOE-H03	Hydraulic Fluids Conditioning and Contamination Control	1.2	12	2	✗	✗	✓	$600

Course Description:

Contamination control is a crucial for hydraulic systems to survive and to sustain their reliability and performance. Hydraulic fluids are inevitably contaminated by various sources. Hydraulic fluid contamination is not limited to just the particulate contaminants as many people may think. Hydraulic fluid contamination can be broadly defined as any internal or external reason that can change the properties or performance. Therefore, this 2-days (12-hour) seminar is designed to cover the knowledge of hydraulic fluids and contamination control. The seminar discusses thoroughly the different types of hydraulic fluids, their properties and standard methods of testing. The seminar also covers all types of contamination, their sources, effects, and best practices to avoid and control them.

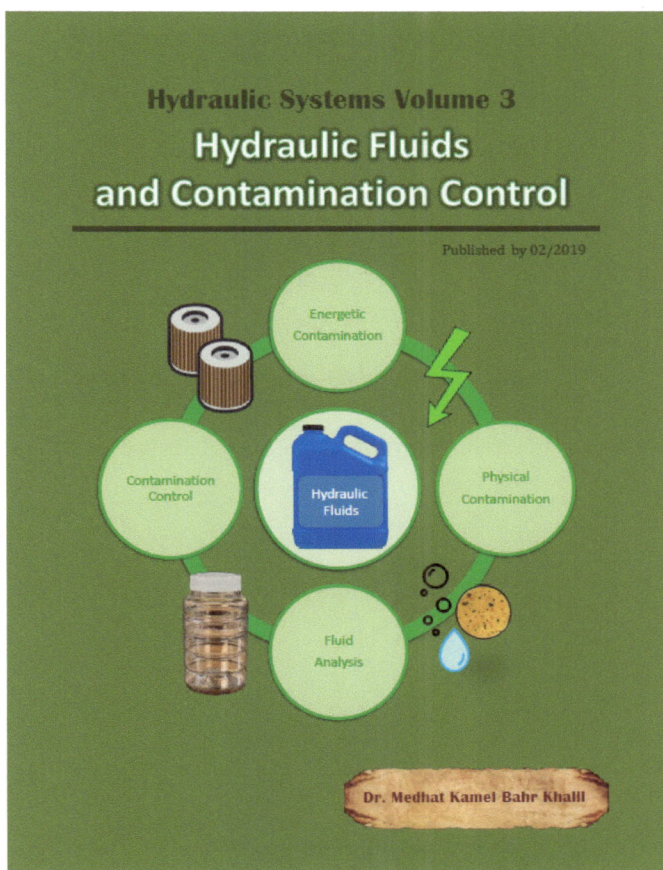

Hydraulic Systems Volume 3
Hydraulic Fluids and Contamination Control
Published by 02/2019

Energetic Contamination · Physical Contamination · Hydraulic Fluids · Fluid Analysis · Contamination Control

Dr. Medhat Kamel Bahr Khalil

ISBN-13: 9780997781632

Course Agenda

Course Agenda: AM Session (9-Noon)   Lunch Hour (Noon - 1 pm)   PM Session (1 - 4)

	Day 1	Hr
AM	**CH01:** Introduction	0.5
	CH02: Hydraulic Fluids	2.5
PM	**CH03:** Energetic Contamination	0.5
	CH04: Gaseous Contamination	0.5
	CH05: Fluidic Contamination	1.0
	CH06: Chemical Contamination	1.0
	Day 2	**Hr**
AM	**CH07:** Particulate Contamination	1.0
	CH08: Hydraulic Fluid Analysis	2.0
PM	**CH09:** Hydraulic Filters Performance Ratings	2.0
	CH10: Contamination Control in Hydraulic Transmission Lines	1.0
	Total	**12**

MSOE-PERD   www.msoe.edu/seminars   khalil@msoe.edu   Cell: +1-414-940-2232

Hydraulic System Modelling and Simulation for Application Engineers

Course #	Course Title	CEU	Hr	Days	Hands-On	Exam	Scheduled	$/Person
	Fluid Power Training							
MSOE-H07	Hydraulic System Modelling and Simulation for Application Engineers	2.7	27	5	✓	✗	✓	$2,160

Course Description:

This 27-hour 5-day presents lumped modeling technique, using Matlab-Simulink, to model discrete hydraulic components that can be recharacterized and used repeatedly in system models. The course applies the lumped modeling concept on hydraulic fluids, transmission lines, pumps, motors, cylinders, pressure relief valves, flow control valves, proportional valves, and servo valves. The course uses the component lumped models to assemble electrohydraulic cylinder position control system and electrohydraulic motor speed control as case studies. The course contains several lab exercises to develop the static characteristics, step response and frequency response of several components. The course also capture the dynamics of control systems to validate the models.

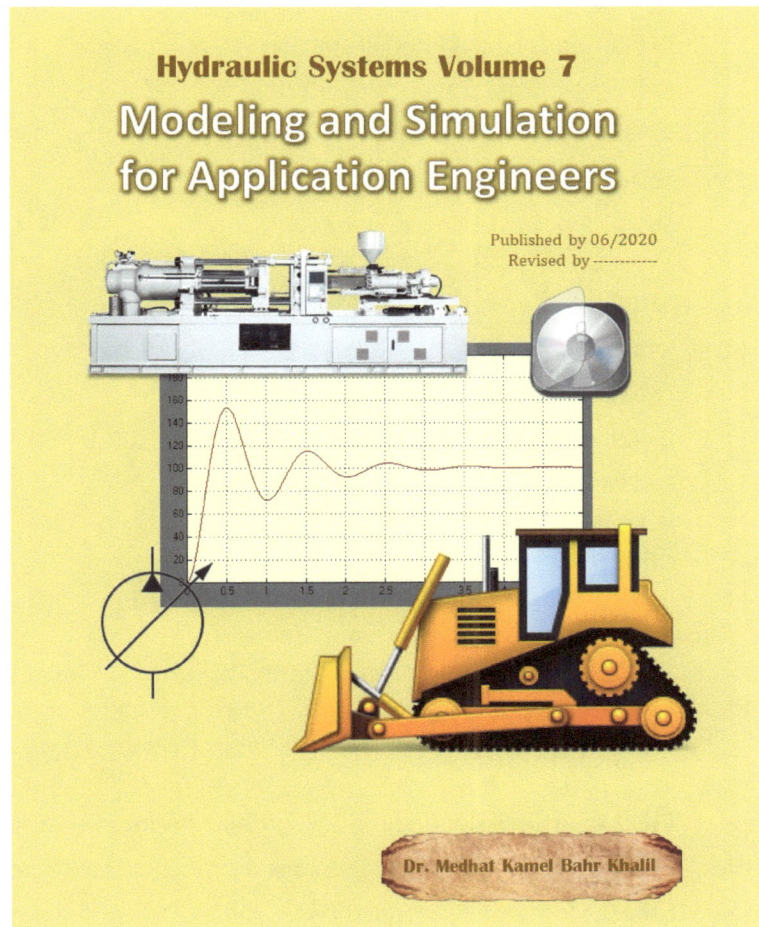

Hydraulic Systems Volume 7
Modeling and Simulation for Application Engineers

Published by 06/2020
Revised by -----------

Dr. Medhat Kamel Bahr Khalil

ISBN: 978-0-9977634-3-0

Course Agenda

Course Agenda: AM Session (9-Noon)   Lunch Hour (Noon - 1 pm)   PM Session (1 - 4)

	Day 1	Hr
AM	Registration and Orientation Session	0.25
	CH01: Introduction to Physical System Modelling and Simulation	0.75
	CH02: Modeling and Simulation of First-Order Dynamic Systems	2
PM	**CH03:** Modeling and Simulation of Second-Order Dynamic Systems	2
	CH04: Hydraulic Components Modeling Approaches	1
	Day 2	**Hr**
AM	**CH05:** Modeling of Fluid Properties	1.5
	CH06: Hydraulic Conductors Modelling	1.5
PM	**CH07:** Modeling of Hydraulic Pumps	2
	Lab09: Pump Static Characteristic Measuring	0.5
	Lab10: Pump Step Response Measuring	0.5
	Day 3	**Hr**
AM	**CH08:** Modeling of Hydraulic Motors	2
	Lab11: Hydraulic Motor U-n Static Characteristics	0.5
	Lab12: Identify Hydraulic Motor Dynamics	0.5
PM	**CH09:** Modelling for Hydraulic Cylinders	2.5
	Lab13: Identify Horizontal Cylinder Dynamics	0.5
	Day 4	**Hr**
AM	**CH10:** Modeling of Hydraulic Valves	2
	Lab14: Proportional Valve Flow Gain Measuring	0.5
	Lab15: Servo Valve Flow Gain Measuring	0.5
PM	**CH11:** Modeling of Hydraulic Control Systems (Cylinder Position Control)	2
	Lab16: EH Position Controlled Hydraulic Cylinder Step Response	0.5
	Lab17: EH Position Controlled Hydraulic Cylinder Frequency Response	0.5
	Day 5	**Hr**
AM	**CH11 (continue):** Modeling of Hydraulic Control Systems (Motor Speed Control)	2
	Lab18: EH Speed Controlled Hydraulic Motor Step Response	0.5
	Lab19: EH Speed Controlled Hydraulic Motor Frequency Response	0.5
	Total	**27**

Design Strategies for Hydraulic Systems

	Fluid Power Training							
Course #	Course Title	CEU	Hr	Days	Hands-On	Exam	Scheduled	$/Person
MSOE-H08	Design Strategies for Hydraulic Systems	2.7	27	5	✓	✗	✓	$2,160

Course Description:

This 27-hour 5-day seminar focuses on the control strategies as applied to hydraulic systems including reviewing basic control theory. The seminar covers various methods of controlling hydraulic systems using open versus closed loop, using PC-Based Control versus PLC-based control. The seminar also covers sensors calibration and the step-by-step calculation for sizing electro-hydraulic systems to meet certain dynamics.

Course Agenda:

Under-development.

Design Strategies for Electro-Hydraulic Systems

		Fluid Power Training						
Course #		Course Title	CEU	Hr	Days	Hands-On	Exam	Scheduled	$/Person
MSOE06		Design Strategies for Electro-Hydraulic Systems	2.7	27	5	✓	✗	UD	$2,160

Course Description:

This 27-hour 5-day seminar focuses on the control strategies as applied to hydraulic systems including reviewing basic control theory. The seminar covers various methods of controlling hydraulic systems using open versus closed loop, using PC-Based Control versus PLC-based control. The seminar also covers sensors calibration and the step-by-step calculation for sizing electro-hydraulic systems to meet certain dynamics.

Course Agenda:

Under-development.

Fundamentals of Hydraulic Systems

Fluid Power Training

Course #	Course Title	CEU	Hr	Days	Hands-On	Exam	Scheduled	$/Person
MSOE-HF01	Fundamentals of Hydraulic Systems	0.5	5	1	✗	✗	✗	$300

Course Description:

This 5-hour 1-day seminar is designed to introduce an overview of hydraulic systems for people who are inexperienced or have limited understanding of hydraulic control technology. The course helps better understanding to the structure and components that form a hydraulic system without deep or specialized discussions and with minimum math contents. The course contains five one-hour parts. The course covers hydraulic symbols and laws of physics for fluid pressure and flow. Construction and principle of operation of basic pumps, motors, valves, and reciprocating actuators are discussed. Simple hydraulic circuits that show how to do basic actions are introduced such as, pump unloading, pressure limiting, motion control of single and multiple actuators, control of overrunning load, speed control of an actuator, and sequence control.

Course Agenda:

Course Outline/Agenda: AM Session (9-Noon) Lunch Hour (Noon - 1 pm) PM Session (1 - 4)

	Day 1	Hr	# of Slides
AM	**Part 1:** Basic Structure and Concepts of Hydraulic Power	1	
	Part 2: Hydraulic Pumps and Motors	1	
	Part 3: Hydraulic Reciprocating Actuators	1	
PM	**Part 4:** Hydraulic Valves	1	
	Part 5: Hydraulic Circuits	1	
	Conclusion	-	

Fundamentals of Electro-Hydraulic Systems

Fluid Power Training

Course #	Course Title	CEU	Hr	Days	Hands-On	Exam	Scheduled	$/Person
MSOE-HF02	Fundamentals of Electro-Hydraulic Systems	0.5	5	1	✗	✗	✗	$300

Course Description:

This 5-hour 1-day seminar is designed to introduce an overview of electro-hydraulic systems for people who are inexperienced or have limited understanding of electro-hydraulic control technology. The course helps better understanding to the structure and components that form an electro-hydraulic system without deep or specialized discussions and with minimum math contents. The course contains five one-hour parts. The course covers electro-hydraulic symbols and laws of physics for electro-magnetic forces. Construction and principle of operation of basic on/off valves, proportional valves, and servo valves are discussed. The course also covers simple electrical circuits that drives the on/off solenoid-operated valves in addition to the control electronics that drives the proportional and servo valves. The course also covers how to select a valve for an open-loop and closed-loop control application.

Course Agenda:

Course Outline/Agenda: AM Session (9-Noon) Lunch Hour (Noon - 1 pm) PM Session (1 - 4)

	Day 1	Hr	# of Slides
AM	**Part 1:** EH Switching Valves	1	
	Part 2: Electrical Circuits for Switching EH Valves	1	
	Part 3: Proportional and Servo Valves	1	
PM	**Part 4:** Open-loop versus Closed-Loop Electro-Hydraulic Control	1	
	Part 5: EH Valve Selection Criteria	1	
	Conclusion	-	

Fundamentals of Service and Operation of Hydraulic Systems

Fluid Power Training

Course #	Course Title	CEU	Hr	Days	Hands-On	Exam	Scheduled	$/Person
MSOE-HF03	Fundamentals of Hydraulic Systems Servicing and Operation	0.5	5	1	✖	✖	✖	$300

Course Description:

This 5-hour 1-day seminar is designed to overview the basic knowledge need to be known by hydraulic system maintenance and troubleshooting personnel in addition to machine operators. The course covers safety precautions, hydraulic fluids, contamination control, hydraulic conductors, accessories, maintenance guidelines, and troubleshooting logics.

Course Agenda:

Course Outline/Agenda: AM Session (9-Noon)   Lunch Hour (Noon - 1 pm)   PM Session (1 - 4)

	Day 1	Hr	# of Slides
AM	**Part 1:** Hydraulic System Safety	1	
	Part 2: Hydraulic Fluids and Contamination Control	1	
	Part 3: Hydraulic Accessories	1	
PM	**Part 4:** Maintenance and Failure Analysis of Hydraulic Components	1	
	Part 5: Hydraulic Systems Troubleshooting	1	
	Conclusion	-	

Overview of Fluid Power Systems

Fluid Power Training

Course #	Course Title	CEU	Hr	Days	Hands-On	Exam	Scheduled	$/Person
WMK01	Overview of Fluid Power Systems	0.6	6	1	✗	✗	✗	$300

Course Description:

This 6-hour one-day course is an introductory overview of fluid power (hydraulic and pneumatic) technology. It is not a scientific study, instead it focuses on hydraulics practiced in the field. In this course, we overview basic principles and components along with simple circuitry. Text book will given to students. You will find many charts of design data at the back of the text book. When you have completed this study, and if you are a serious student, you will want to continue your study with the more advanced courses.

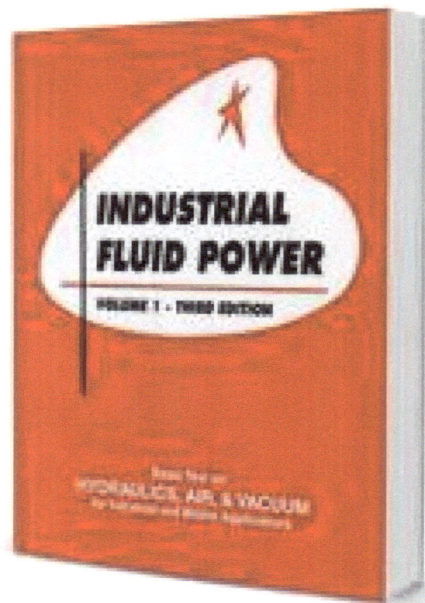

Course Agenda:

Course Outline/Agenda: AM Session (9-Noon)   Lunch Hour (Noon - 1 pm)   PM Session (1 – 4)

	Day 1	Hr	# of Slides
AM	**CH01:** Fluid Power Principles	1	46
	CH02: Fluid Power Cylinders	1	78
	CH03: Fluid Power Valves I	1	64
PM	**CH04:** Fluid Power Valves II	1	46
	CH05: Fluid Power Pumps	1	47
	CH06: Fluid Power Accessories	1	40
	Total Contact Hours	6	

Fluid Power Applications

		Fluid Power Training						
Course #	Course Title	CEU	Hr	Days	Hands-On	Exam	Scheduled	$/Person
WMK02	Fluid Power Applications	1.2	12	2	✗	✗	✗	$600

Course Description:

This 12-hour 2-day course covers fluid power (hydraulic and pneumatic) application. Topics include circuits to perform, pressure control, speed control and general actuator motion actuator (hydraulic and pneumatic). Text book will be given to students. When you have completed this study, and if you are a serious student, you will want to continue your study with more advanced courses.

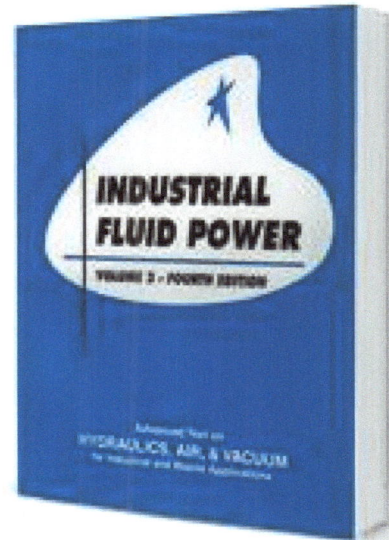

Course Agenda:

Course Outline/Agenda: AM Session (9-Noon) Lunch Hour (Noon - 1 pm) PM Session (1 - 4)

	Day 1	**Hr**	**# of Slides**
AM	**CH01:** Fluid Power Cylinders	1	27
	CH02: Introduction To Air Circuitry	1	32
	CH03: Several Cylinders On One Machine	1	26
PM	**CH04:** Automatic Reciprocation	1	13
	CH05: Miscellaneous Air Circuits	1	14
	CH06: Hydraulic Circuitry	1	9
	Day 2	**Hr**	**# of Slides**
AM	**CH07:** Directional Control	1.5	65
	CH08: Pressure Control	1.5	48
PM	**CH09:** Speed Control	0.75	15
	CH10: Hydraulic Circuit	0.75	14
	CH11: Pressure Intensification	0.75	18
	CH12: Air Over Oil Applications	0.75	12
	Total Contact Hours	**12**	

Hydraulic Motors Construction and Operation

Fluid Power Training

Course #	Course Title	CEU	Hr	Days	Hands-On	Exam	Scheduled	$/Person
WMK03	Hydraulic Motors Construction and Application	1.2	12	2	✗	✗	✗	$600

Course Description:

This 12-hour 2-day course covers fluid power rotational actuators (hydraulic and pneumatic) construction and application technology. It is not a scientific study, instead it covers practices used in the field. In this course, we overview various constructions of air and hydraulic motors and samples of circuits and applications using rotational motion. Text book will given to students. You will find many charts of design data at the back of the text book. When you have completed this study, and if you are a serious student, you will want to continue your study with more advanced courses.

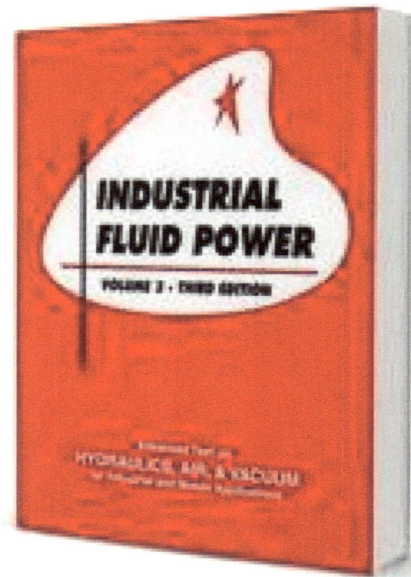

Course Agenda:

Course Outline/Agenda: AM Session (9-Noon)   Lunch Hour (Noon - 1 pm)   PM Session (1 - 4)

	Day 1	Hr	# of Slides
AM	**CH01:** Hydraulic Motors Compared To Hydraulic Pumps	1	16
	CH02: A Review Of Force	1	19
	CH03: Torque Determination	1	13
PM	**CH04: Hydraulic and Electric Motors Compared**	0.5	7
	CH05: Motor Circuits For One Direction of Rotation	1.5	30
	CH06: Basic Methods Of Motor Speed Control	1	27
	Day 2	Hr	# of Slides
AM	**CH07: Side Loading On Motor Shaft**	1	10
	CH08: The Closed Loop For Transmitting Power	1	23
	CH09: Air Motor Types	1	10
PM	**CH10:** Rotary Actuators	1	12
	CH11: Rotary-Type Flow Dividers	1	22
	CH12: Types Of Steering Linkage	0.5	5
	CH13: Tension Stressing	0.5	13
	Total Contact Hours	12	

Pneumatic Specialist Certification Review Session

Course #	Course Title	CEU	Hr	Days	Hands-On	Exam	Scheduled	$/Person
MSOE00-P	Pneumatic Specialist Certification Review Session (IFPS Certification)	1.8	18	3	✗	✓	✗	$600

Fluid Power Training

Course Description:

This 18-hours 3-days review session is conducted at MSOE followed by the certification exam on the fourth day. The objective of the course is to walk the participants through the study manual provided by IFPS in order to maximize the chance of passing the certification exam.

International **Fluid Power**™ Society

What is the IFPS?

The International Fluid Power Society is the only organization that provides comprehensive technical certification offerings for all professionals in the fluid power and motion control industry.

What is the Process of Certification?

After 3-days review session provided by MSOE, participants will take the certification exam on forth day. Exam will be provided and proctored by IFPS. The test is 3-hours 50-questions multiple-choice type of test. You need to get 35 correct answers out of 50 questions. If you fail you can re-schedule taking the exam at a later time. If you pass, you will be issued a "Pneumatic Specialist" certificate. The certificate is good for five years, after five years you do not need to retake the exam, you need only to report to IFPS indicating that you are still working in the field.

Why Get Certified?

- The "Pneumatic Specialist" certification is an internationally recognized certification.
- The certificate is portable - it goes with the individual wherever they work.
- Certifications help an individual to advance his career and introduce himself to the global job market.
- Certification sets an individual apart as a leader in their chosen field of work.
- Certification will help a vendor provide over-the-top quality and acquire ISO certification easily.
- Certified personnel help make the work environment safe and improve the safety, reliability and efficiency of a machine operation.

MSOE-PERD   www.msoe.edu/seminars   khalil@msoe.edu   Cell: +1-414-940-2232

Course Agenda

Course Agenda: AM Session (9-Noon)   Lunch Hour (Noon - 1 pm)   PM Session (1 - 4)

	Day 1	Hr
AM	Registration and Orientation Session	0.5
	Job Responsibility 1: Load and Motion Analysis	2.5
PM	**Job Responsibility 2:** System Analysis and Troubleshooting	1.5
	Job Responsibility 3: System Design	1.5
	Day 2	**Hr**
AM	**Job Responsibility 3:** Contd.	1.0
	Job Responsibility 4: Component Application	2.0
PM	**Job Responsibility 5:** Air Compress ion and Preparation	1.5
	Job Responsibility 6: Control Components and Systems	1.5
	Day 3	
AM	**Pretest 1 and 2**	3.0
PM	**Pretest 3 and 4**	3.0
	Total	**18**
	Day 4	
AM	**Certification Exam**	3

Introduction to Pneumatic Systems for Application Engineers

Fluid Power Training

Course #	Course Title	CEU	Hr	Days	Hands-On	Exam	Scheduled	$/Person
MSOE07	Introduction to Pneumatic Systems for Application Engineers	1.8	18	3	✓	✗	UD	$600

Course Description:

This 18-hour 3-day seminar is designed to cover pneumatic systems in-depth, including design concepts and calculations. The seminar covers sizing calculations, read schematics, mechanical valves, actuators, compressors and air treatment.

Course Agenda:

Under-development.

Electro-Pneumatic Components and Systems

Fluid Power Training

Course #	Course Title	CEU	Hr	Days	Hands-On	Exam	Scheduled	$/Person
MSOE08	Electro-Pneumatic Components and Systems	1.8	18	3	✓	✗	UD	$600

Course Description:

This 18-hour 3-day seminar is designed to cover in-depth electro-pneumatics including components and systems. The seminar covers solenoid and proportional valves, pneumatic accessories used in industrial automation and how to use pneumatic devices to build open and closed loop control systems.

Course Agenda:

Under-development.

Designation Table

Condition	Des.	Clarification
Exam:	✓	Course contains certification exam to get certified
	✗	No certification exam.
Hands-On:	✓	Course contains hands-on labs.
	✗	Course conducted on theoretical base.
Scheduled:	✓	Course scheduled and registration is opened for public
	✗	Course is offered upon request at the customer-site or for public when the minimum enrollment number is reached.
	UD	Course is under development.

Customize Your Own Industrial Training.
Courses can be mobilized to your facility.
Courses in this sectors are non-scheduled courses offered only in customer-site.
If there is an interest, please contact Dr. Medhat Khalil directly.

Mechanical Systems Training

Course #	Course Title	CEU	Hr	Days	Hands-On	Exam	Scheduled	$/Person
ATP02	Industrial Mechanics	2.7	27	5	✗	✓	✗	$1,350
ATP03	Small Engines	2.1	21	4	✗	✓	✗	$1,050
ATP04	Low Pressure Boilers	1.8	18	3	✗	✓	✗	$900
ATP05	High Pressure Boilers	1.8	18	3	✗	✓	✗	$900

Industrial Mechanics

Mechanical Systems Training

Course #	Course Title	CEU	Hr	Days	Hands-On	Exam	Scheduled	$/Person
ATP02	Industrial Mechanics	2.7	27	5	✗	✓	✗	$1,350

Course Description:

This 27-hour 5-day seminar presents an overview of the principles of industrial mechanical systems and the equipment in these systems. This seminar presents the latest information on all aspects of mechanical systems, including rigging, lifting, ladders and scaffolds, hydraulic systems, pneumatic systems, lubrication, bearings, belts and pulleys, mechanical drives, vibration, alignment, and electricity. Industrial Mechanics is designed for postsecondary, industrial, and apprenticeship training programs. To assure high level of contribution of the participants, each chapter concludes with self assessment test. Certification will be granted only for people who pass the final Certification Exam. Post test will be retaken by people who failed on first one. A CD-ROM is included and contains information to supplement the textbook.

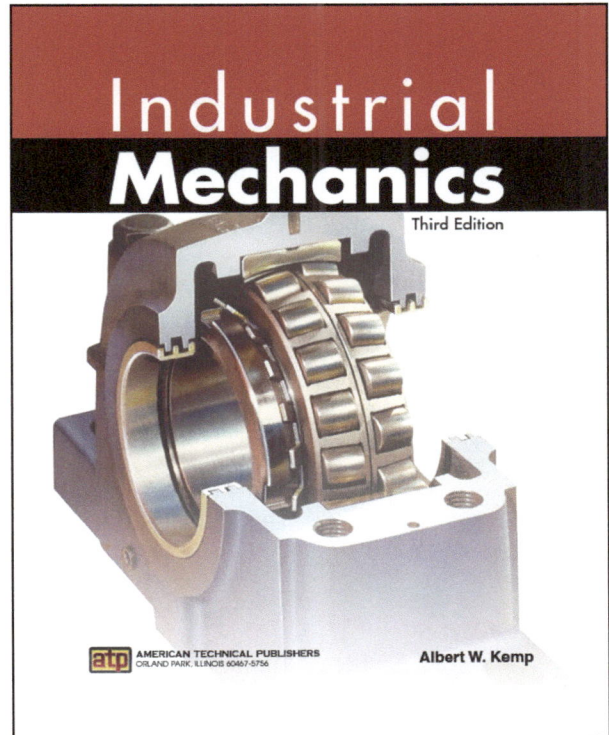

Course Agenda

Course Outline/Agenda: AM Session (9-Noon)   Lunch Hour (Noon - 1 pm)   PM Session (1 - 4)

	Day 1	Hr	# of Slides
AM	**Pre-Test**	2	0
AM	**CH01:** Industrial Safety	1	28
PM	**CH02:** Precision Measurement	1.5	44
PM	**CH03:** Print reading	1.5	46
	Day 2	Hr	
AM	**CH04:** Tools (self Study)	0	28
AM	**CH05:** Calculations	1	21
AM	**CH06:** Rigging (self Study)	0	64
AM	**CH07:** Lifting	1	36
AM	**CH08:** Ladders and Scaffolds (Self Study)	0	29
AM	**CH09:** Hydraulic Principles	1	30
PM	**CH10:** Hydraulic Applications	3	60
	Day 3	Hr	
AM	**CH11:** Pneumatic Principles	1	21
AM	**CH12:** Pneumatic Applications	1	35
AM	**CH13:** Lubrication	1	16
PM	**CH14:** Bearings	1	22
PM	**CH15:** Flexible Belt Drives	1	22
PM	**CH16:** Mechanical Drives	1	18
	Day 4	Hr	
AM	**CH17:** Vibration	1	24
AM	**CH18:** Alignment	1	30
	CH19: Electricity	1.5	33
	CH20: Preventive Maintenance Programs	1.5	24
	Day 5	Hr	
AM	**Review and conclusion**	1	
AM	**Certification Exam**	2	
	Total	27	

MSOE-PERD   www.msoe.edu/seminars   khalil@msoe.edu   Cell: +1-414-940-2232

Small Engines

Mechanical Systems Training

Course #	Course Title	CEU	Hr	Days	Hands-On	Exam	Scheduled	$/Person
ATP03	Small Engines	2.1	21	4	✗	✓	✗	$1,050

Course Description:

This 21-hour 4-day seminar presents Small Engines in a comprehensive textbook that presents small engine operation and service principles using concise text, detailed illustrations, and practical applications. The content is based on technician requirements put forth by Briggs & Stratton. The seminar explains the why of engine design and the how of operation as well as basic repair. A CD-ROM is included and contains information to supplement the textbook.

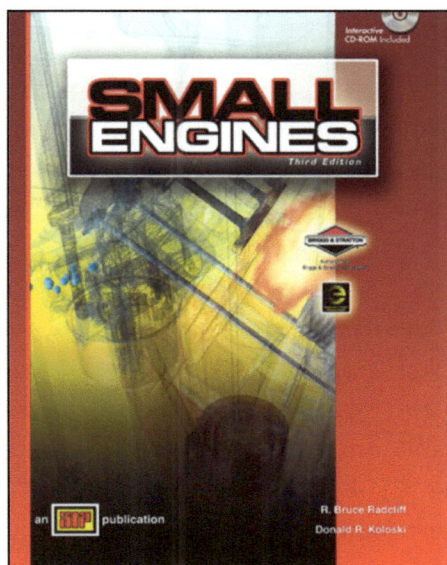

Course Agenda:

Course Outline/Agenda: AM Session (9-Noon)   Lunch Hour (Noon - 1 pm)   PM Session (1 - 4)

	Day 1	Hr	# of Slides
AM	**Pre-Test**	2	0
	CH01: Internal Combustion Engine	1	25
PM	**CH02:** Safety & Tools	1.5	25
	CH03: Engine Operation	1.5	28
	Day 2	**Hr**	
AM	**CH04:** Compression System	1.5	36
	CH05: Fuel System	1.5	39
PM	**CH06:** Governor System	1	26
	CH07: Electrical System	2	41
	Day 3	**Hr**	
AM	**CH08:** Cooling and Lubrication System	1	29
	CH09: Multiple Cylinder Engines	1	18
	CH10: Troubleshooting	1	32
PM	**CH11:** Failure Analysis	1	24
	CH12: Engine Application and Selection	2	49
	Day 4	**Hr**	
AM	**Review and conclusion**	1	
	Certification Exam	2	
	Total	**21**	

MSOE-PERD   www.msoe.edu/seminars   khalil@msoe.edu   Cell: +1-414-940-2232

Low Pressure Boilers

Mechanical Systems Training

Course #	Course Title	CEU	Hr	Days	Hands-On	Exam	Scheduled	$/Person
ATP04	Low Pressure Boilers	1.8	18	3	✗	✓	✗	$900

Course Description:

This 18-hour 3-day seminar presents information on the safe and efficient operation of low pressure steam boilers and related equipment, hot water boilers, and cooling systems. The provided textbook covers the new ASME symbol stamps, integrated boiler controls, code requirements for bottom blow down, feed water regulators, emissions regulations and New Source Performance Standards, variable-speed drives, diaphragm draft gauges, water treatment programs and solubilizing water treatments. Energy efficiency and environmental issues are emphasized throughout. A CD-ROM is included and contains information to supplement the textbook.

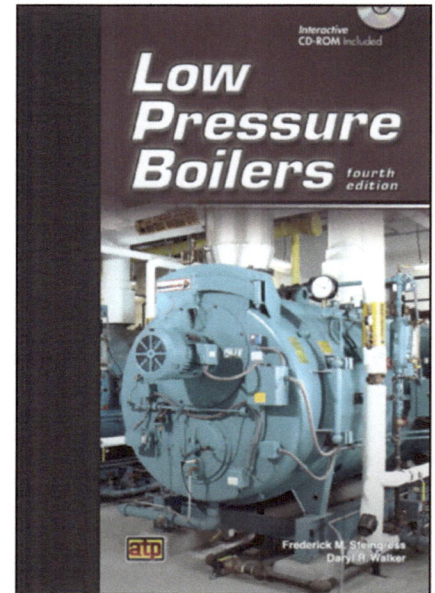

Course Agenda:

Course Outline/Agenda: AM Session (9-Noon) Lunch Hour (Noon - 1 pm) PM Session (1 - 4)

		Hr	# of Slides
	Day 1		
AM	**Pre-Test**	2	0
	CH01: Boiler Operation Principles	1	26
PM	**CH02:** Boiler Fittings	1	24
	CH03: Feed water Systems	1	20
	CH04: Steam System Accessories	1	14
	Day 2		
AM	**CH05:** Fuel Systems	1	32
	CH06: Draft System	1	9
	CH07: Boiler Water Treatment	1	21
PM	**CH08:** Boiler Operation Procedures	1	22
	CH09: Hot Water Heating Systems	1	26
	CH10: Cooling Systems	1	28
	Day 3		
AM	**CH11:** Boiler Operation Safety	1.5	20
	CH12: Boiler Operator Licensing	1.5	29
PM	**Review and conclusion**	1	0
	Certification Exam	2	0
	Total	18	

High Pressure Boilers

Mechanical Systems Training

Course #	Course Title	CEU	Hr	Days	Hands-On	Exam	Scheduled	$/Person
ATP05	High Pressure Boilers	1.8	18	3	✗	✓	✗	$900

Course Description:

This 18-hour 3-day seminar presents provides a comprehensive overview of the safe and efficient operation of high pressure boilers and related equipment. The latest combustion control technology, as well as EPA regulations and their implications, are included in this seminar. This edition has been reorganized to provide a systems view of boiler operation. All aspects of high pressure boilers are discussed and illustrated. The provided text book contains comprehensive glossary and appendix provide helpful reference material. This textbook is designed for both learners preparing to obtain a boiler operator's license and for boiler operators intending to upgrade their skills. A CD-ROM is included and contains information to supplement the textbook.

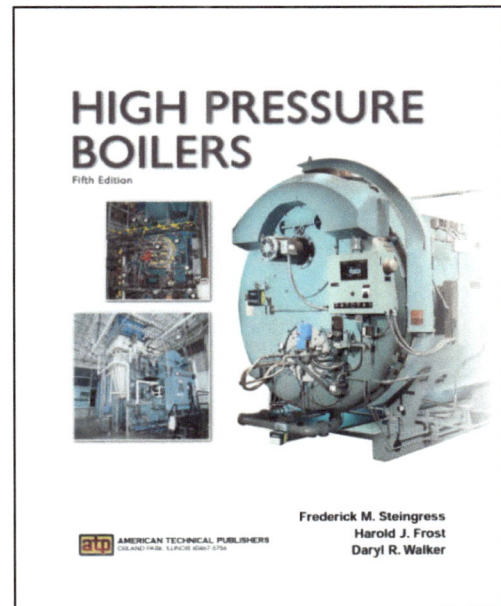

HIGH PRESSURE BOILERS
Fifth Edition

AMERICAN TECHNICAL PUBLISHERS
OLLAND PARK, ILLINOIS 60467 5756

Frederick M. Steingress
Harold J. Frost
Daryl R. Walker

Course Agenda:

Course Outline/Agenda: AM Session (9-Noon) Lunch Hour (Noon - 1 pm) PM Session (1 - 4)			
	Day 1	Hr	# of Slides
AM	Pre-Test	1.5	0
	CH01: Steam Boilers	1.5	35
PM	CH02: Boiler Systems	1	14
	CH03: Steam System Fittings	1	21
	CH04: Steam System Accessories	1	25
	Day 2	Hr	
AM	CH05: Feed-Water Systems	1	24
	CH06: Water Treatment	1	28
	CH07: Combustion Equipment	1	27
PM	CH08: Fuels and Combustion	1	11
	CH09: Combustion and Boiler Controls	1	23
	CH10: Draft Systems	1	17
	Day 3	Hr	
AM	CH11: Instrumentation and Control Systems	1.5	38
	CH12: Steam Boiler Operation	1.5	43
PM	Review and conclusion	1	0
	Certification Exam	2	0
	Total	18	

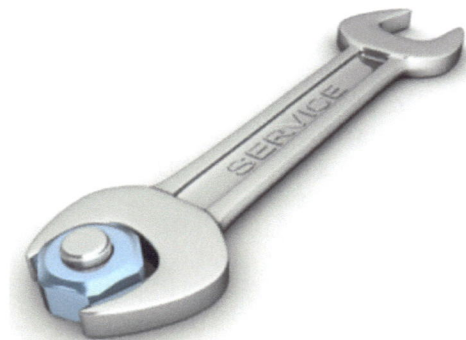

Designation Table

Condition	Des.	Clarification
Exam:	✓	Course contains certification exam to get certified
	✗	No certification exam.
Hands-On:	✓	Course contains hands-on labs.
	✗	Course conducted on theoretical base.
Scheduled:	✓	Course scheduled and registration is opened for public
	✗	Course is offered upon request at the customer-site or for public when the minimum enrollment number is reached.
	UD	Course is under development.

Customize Your Own Industrial Training.
Courses can be mobilized to your facility.
Courses in this sectors are non-scheduled courses offered only in customer-site.
If there is an interest, please contact Dr. Medhat Khalil directly.

Mechanical Maintenance

Course #	Course Title	CEU	Hr	Days	Hands-On	Exam	Scheduled	$/Person
ATP06	Industrial Maintenance	2.1	21	4	✗	✓	✗	$1,050

Industrial Maintenance

Mechanical Maintenance

Course #	Course Title	CEU	Hr	Days	Hands-On	Exam	Scheduled	$/Person
ATP06	Industrial Maintenance	2.1	21	4	✗	✓	✗	$1,050

Course Description:

This 21-hour 4-day emphasizes on maintenance personnel versatility. Industrial Maintenance is a comprehensive source of fundamental system operation, maintenance, and troubleshooting information. This edition builds on industry-proven content and offers expanded coverage in the areas of energy efficiency and auditing, waste reduction, safety standards, advanced multimeter functions and procedures, building automation systems, and indoor air quality. Real-world maintenance problems and solutions are depicted throughout the textbook, along with equipment operating principles, maintenance management procedures, and troubleshooting scenarios for common systems. A CD-ROM is included and contains information to supplement the textbook.

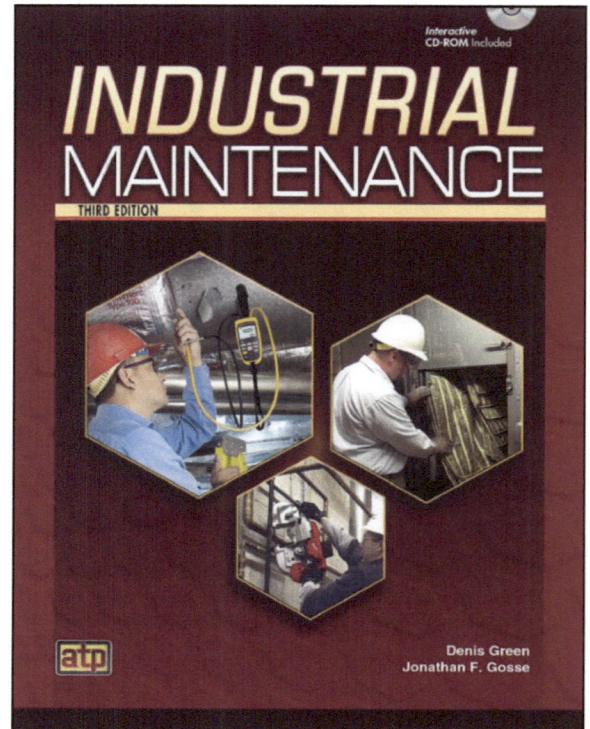

Course Agenda

Course Outline/Agenda: AM Session (9-Noon)   Lunch Hour (Noon - 1 pm)   PM Session (1 - 4)

		Day 1	Hr	# of Slides
AM		**Section 1: Introduction to Instrumentation**		
		CH01: Instrumentation Overview	1	7
		CH02: Fundamentals of Process Control	1	12
		CH03: Piping and Instrumentation Diagram	1	5
PM		**Section 2: Temperature Measurement**		
		CH04: Temperature Heat and Energy	0.75	9
		CH05: Thermal Expansion Thermometer	0.75	10
		CH06: Electrical Thermometers	1.5	24
		Day 2	**Hr**	
AM		**CH07:** Infrared Red Radiation Thermometer	1	15
		CH08: Practical Temperature Measurement and Calibration	2	21
PM		**Section 3: Pressure Measurement**		
		CH09: Electrical Pressure Elements	0.75	9
		CH10: Practical Pressure Measurement and Calibration	0.75	11
		CH11: Mechanical Level Instruments	0.75	9
		CH12: Electrical Level Instruments	0.75	13
		Day 3	**Hr**	
AM		**Section 4: Level Measurement**		
		CH13: Mechanical Level Instruments	0.75	17
		CH14: Electrical Level Instruments	0.75	12
PM		**CH15:** Ultrasonic Radar and Laser Level Instruments	1	8
		CH16: Nuclear level Instruments and Weigh Systems	1	6
		CH17: Practical Level Measurement and Calibration	1	11
		Day 4	**Hr**	
AM		**Section 5: Flow Measurement**		
		CH18: Fluid Flow	1	7
		CH19: Differential Pressure Flow meter	1	8
		CH20: Mechanical Flow meter	1	11
PM		**CH21:** Magnetic Ultrasonic and Mass Flow meter	0.75	8
		CH22: Practical Flow Measurement	0.75	6
		Section 6: Analyzers		
		CH23: Gas Analyzer	0.75	12
		CH24: Humidity and Solid Moisture Analyzer	0.75	11
		Day 5	**Hr**	
AM		**CH25:** Liquid Analysers	0.75	13
		CH26: Electromechanically and Composition Analyzer	0.75	17
		Section 7: Position Measurement		
		CH27: Mechanical and Proximity Switch	0.75	13
		CH28: Practical Position Measurement	0.75	10
		Total	**27**	

Process Engineering Training

60

Designation Table

Condition	Des.	Clarification
Exam:	✓	Course contains certification exam to get certified
	✗	No certification exam.
Hands-On:	✓	Course contains hands-on labs.
	✗	Course conducted on theoretical base.
Scheduled:	✓	Course scheduled and registration is opened for public
	✗	Course is offered upon request at the customer-site or for public when the minimum enrollment number is reached.
	UD	Course is under development.

Customize Your Own Industrial Training.
Courses can be mobilized to your facility.
Courses in this sectors are non-scheduled courses offered only in customer-site.
If there is an interest, please contact Dr. Medhat Khalil directly.

Process Engineering

Course #	Course Title	CEU	Hr	Days	Hands-On	Exam	Scheduled	$/Person
ATP07A	Instrumentation I	2.7	27	5	✗	✓	✗	$1,350
ATP07B	Instrumentation II	2.7	27	5	✗	✓	✗	$1,350
WMK05	Fundamentals of Industrial Control and Automation	1.2	12	2	✗	✓	✗	$600
TPC 271	Introduction to Process Control	0.6	6	1	✗	✓	✗	$300
TPC 272	Foundations of Measurement Instrumentation	0.5	5	1	✗	✓	✗	$250
TPC 273	Pressure Measurement	0.5	5	1	✗	✓	✗	$250
TPC 274	Force, Weight, and Motion Measurement	0.5	5	1	✗	✓	✗	$250
TPC 275	Flow Measurement	1.0	10	2	✗	✓	✗	$500
TPC 276	Level Measurement	0.5	5	1	✗	✓	✗	$250
TPC 281	Working with Controllers	0.5	5	1	✗	✓	✗	$250
TPC 282	How Control Loops Operate	0.5	5	1	✗	✓	✗	$250

Instrumentation I

Process Engineering

Course #	Course Title	CEU	Hr	Days	Hands-On	Exam	Scheduled	$/Person
ATP07A	Instrumentation I	2.7	27	5	✗	✓	✗	$1,350

Course Description:

This 27-Hours 5-Day seminar is PART 1 of a comprehensive review that provides a technician-level approach to instrumentation used in process control. With an emphasis on common industrial applications, this textbook covers the four fundamental instrumentation measurements of temperature, pressure, level, and flow, in addition to position, humidity, moisture, and typical liquid and gas measuring instruments. Fundamental scientific principles, detailed illustrations, descriptive photographs, and concise text are used to present the following instrumentation topics. A CD-ROM is included and contains information to supplement the textbook.

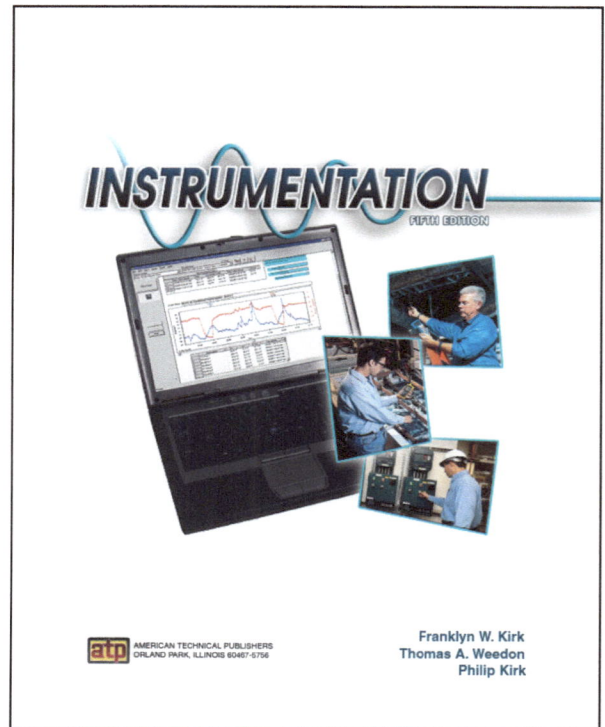

INSTRUMENTATION
FIFTH EDITION

atp AMERICAN TECHNICAL PUBLISHERS
ORLAND PARK, ILLINOIS 60467-5756

Franklyn W. Kirk
Thomas A. Weedon
Philip Kirk

Process Engineering Training

Course Agenda

Course Outline/Agenda: AM Session (9-Noon)   Lunch Hour (Noon - 1 pm)   PM Session (1 - 4)

	Day 1	Hr	# of Slides
	Section 1: Introduction to Instrumentation		
AM	**CH01:** Instrumentation Overview	1	7
	CH02: Fundamentals of Process Control	1	12
	CH03: Piping and Instrumentation Diagram	1	5
	Section 2: Temperature Measurement		
PM	**CH04:** Temperature Heat and Energy	0.75	9
	CH05: Thermal Expansion Thermometer	0.75	10
	CH06: Electrical Thermometers	1.5	24
	Day 2	**Hr**	
AM	**CH07:** Infrared Red Radiation Thermometer	1	15
	CH08: Practical Temperature Measurement and Calibration	2	21
	Section 3: Pressure Measurement		
	CH09: Electrical Pressure Elements	0.75	9
PM	**CH10:** Practical Pressure Measurement and Calibration	0.75	11
	CH11: Mechanical Level Instruments	0.75	9
	CH12: Electrical Level Instruments	0.75	13
	Day 3	**Hr**	
	Section 4: Level Measurement		
AM	**CH13:** Mechanical Level Instruments	0.75	17
	CH14: Electrical Level Instruments	0.75	12
	CH15: Ultrasonic Radar and Laser Level Instruments	1	8
PM	**CH16:** Nuclear level Instruments and Weigh Systems	1	6
	CH17: Practical Level Measurement and Calibration	1	11
	Day 4	**Hr**	
	Section 5: Flow Measurement		
AM	**CH18:** Fluid Flow	1	7
	CH19: Differential Pressure Flow meter	1	8
	CH20: Mechanical Flow meter	1	11
	CH21: Magnetic Ultrasonic and Mass Flow meter	0.75	8
	CH22: Practical Flow Measurement	0.75	6
PM	**Section 6: Analysers**		
	CH23: Gas Analyzer	0.75	12
	CH24: Humidity and Solid Moisture Analyzer	0.75	11
	Day 5	**Hr**	
	CH25: Liquid Analysers	0.75	13
	CH26: Electromechanically and Composition Analyzer	0.75	17
AM	**Section 7: Position Measurement**		
	CH27: Mechanical and Proximity Switch	0.75	13
	CH28: Practical Position Measurement	0.75	10
	Total	27	

Instrumentation II

Process Engineering

Course #	Course Title	CEU	Hr	Days	Hands-On	Exam	Scheduled	$/Person
ATP07B	Instrumentation II	2.7	27	5	✗	✓	✗	$1,350

Course Description:

This 27-Hours 5-Day seminar is PART 2 of a comprehensive review that provides a technician-level approach to instrumentation used in process control. With an emphasis on common industrial applications, this textbook covers the four fundamental instrumentation measurements of temperature, pressure, level, and flow, in addition to position, humidity, moisture, and typical liquid and gas measuring instruments. Fundamental scientific principles, detailed illustrations, descriptive photographs, and concise text are used to present the following instrumentation topics. A CD-ROM is included and contains information to supplement the textbook.

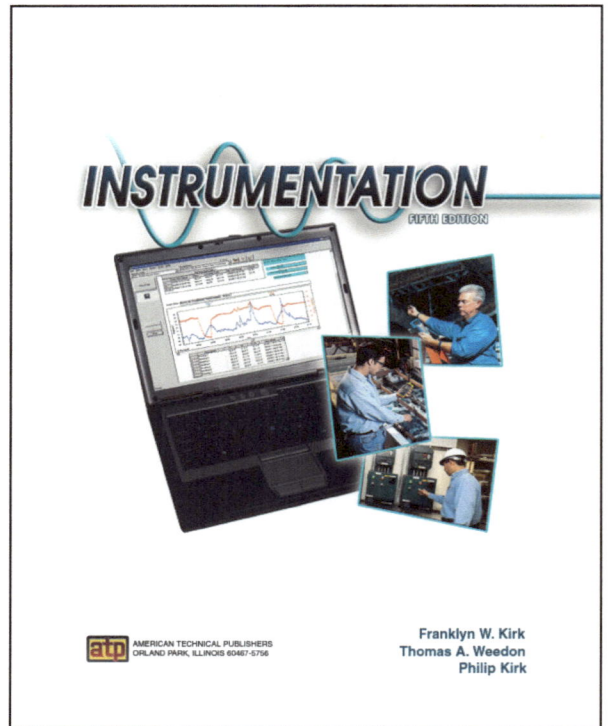

INSTRUMENTATION
FIFTH EDITION

AMERICAN TECHNICAL PUBLISHERS
ORLAND PARK, ILLINOIS 60467-5756

Franklyn W. Kirk
Thomas A. Weedon
Philip Kirk

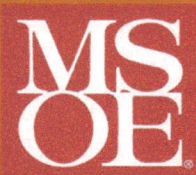

Course Agenda

Course Outline/Agenda: AM Session (9-Noon)   Lunch Hour (Noon - 1 pm)   PM Session (1 - 4)

	Day 1	Hr	# of Slides
	Section 8: Transmission and Communication		
AM	**CH29:** Transmission Signal	1	7
	CH30: Digital Numbering System and Codes	1	9
	CH31: Digital Communications	1	14
	CH32: Industrial Networks	1	16
	CH33: Wireless Systems	1	12
	CH34: Practical Transmission and Communication	1	8
	Day 2	**Hr**	
	Section 9: Automatic Control		
AM	**CH35:** Automatic Control and Process Dynamics	1	19
	CH36: Control Strategies	2	20
	CH37: Control Tuning	0.75	10
PM	**CH38:** Digital and Electric Controllers	0.75	15
	Section 10: Final Elements		
	CH39: Control Valves	1.5	21
	Day 3	**Hr**	
	CH40: Regulators and Dampers	0.75	10
AM	**CH41:** Actuators and Positioners	1.5	20
	CH42: Variable Speed Drives and Electric Power Controllers	0.75	9
	Section 11: Safety Systems		
PM	**CH43:** Safety Devices and Equipment	1.5	21
	CH44: Electrical Safety Standards	0.75	8
	CH45: Safety Instrumented Systems	0.75	5
	Day 4	**Hr**	
	Section 12: Instrumentation and Control Applications		
AM	**CH46:** General Control Techniques	2	19
	CH47: Temperature Control	1	8
	CH48: Pressure and Level Control	1	6
PM	**CH49:** Flow Control	1	12
	CH50: Analysis and Multi-Variable Control	1	6
	Day 5	**Hr**	
AM	**Review and Conclusion**	1	
	Certification Exam	2	
	Total	**27**	

Fundamentals of Industrial Control and Automation

Process Engineering

Course #	Course Title	CEU	Hr	Days	Hands-On	Exam	Scheduled	$/Person
WOMK05	Fundamentals of Industrial Control and Automation	1.2	12	2	✗	✓	✗	$960

Course Description:

This 12-hour 2-day course is an introduction to the Fundamentals of Industrial Controls and Automation. The seminar focuses on a beginner's study of electricity, electronics, control components and automation as related to industrial controls. Modern manufacturing techniques are only possible because of dependable electrical control systems. The provided textbook explores the proper use of electrical controls to maximize productivity, minimize downtime, simplify maintenance, improve safety and provide information to effectively manage operations. Topics covered in the seven chapters include: Electrical Fundamentals, Input Devices - Sensors and Switches, Logic Devices - Timers and Counters, Output Devices, Schematic Diagrams and Logic, Programmable Logic Controllers and Accessories, and Temperature Control Systems.

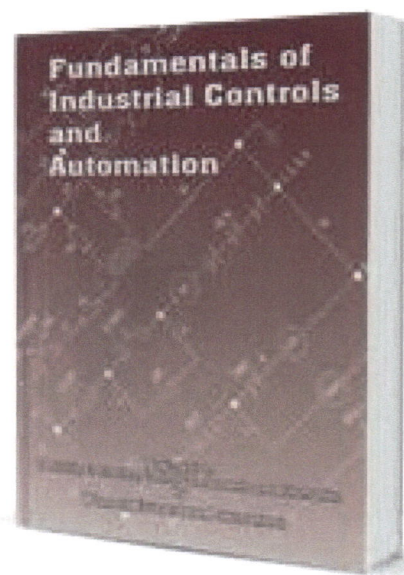

Course Agenda

	Course Outline/Agenda: AM Session (9-Noon) Lunch Hour (Noon - 1 pm) PM Session (1 - 4)		
	Day 1	**Hr**	**# of Slides**
AM	**CH01:** Popular Fluid and Electrical Components	1.5	15
	CH02: How to Draw and Read Electrical Diagrams	1.5	41
PM	**CH03:** Directional Control; Reciprocation of Cylinders	1	19
	CH04: Directional Control; Sequencing of Cylinders	2	40
	Day 2	**Hr**	**# of Slides**
AM + PM	**CH05:** Pressure Control by Electrical Means	1.5	33
	CH06: Solving Design Problems in Electrical Circuitry	3	50
	CH07: Miscellaneous Applications	1.5	21
	Total	12	

Introduction to Process Control

Process Engineering

Course #	Course Title	CEU	Hr	Days	Hands-On	Exam	Scheduled	$/Person
TPC 271	Introduction to Process Control	0.6	6	1	✗	✓	✗	$300

Course Description:

This 6-hours 2-day seminar covers the function of basic devices for measuring and controlling different kinds of variables in process control. Introduces closed-loop control and PID functions. Introduces analog and digital devices and programmable logic controllers (PLCs). ISA and SAMA instrumentation symbols and interpretation and use of process diagrams are covered.

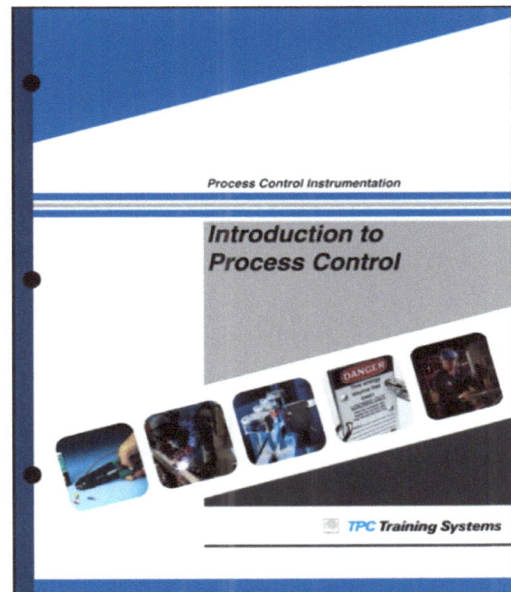

Course Agenda:

Course Outline/Agenda: AM Session (9-Noon) Lunch Hour (Noon - 1 pm) PM Session (1 - 4)			
	Day 1	Hr	# of Slides
AM	**CH01:** The Nature of Process Control	1	
	CH02: Elements of Process Control	1	
	CH03: Process Control Signals	1	
PM	**CH04:** Process Control Diagrams	1	
	CH05: Using Symbols and Diagrams in Process Control	1	
	CH06: Process Control Loop Operations	1	
	Total Contact Hours	6	
	Certification Exam	2	

Foundations of Measurement Instrumentation

Process Engineering

Course #	Course Title	CEU	Hr	Days	Hands-On	Exam	Scheduled	$/Person
TPC 272	Foundations of Measurement Instrumentation	0.5	5	1	✗	✓	✗	$250

Course Description:

This 5-hours 1-day course covers the basic principles of measurement and defines process control terms. Describes several kinds of signals and displays and traces the path of a signal through the system. Explains the operation of transducers, transmitters, signal conditioners, converters, and recorders. Discusses specification details, conversion between English and SI units, calibration methods, and the maintenance of records.

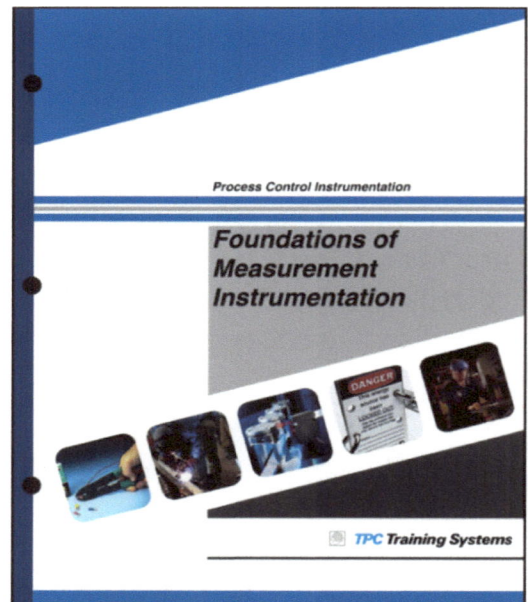

Course Agenda:

Course Outline/Agenda: AM Session (9-Noon) Lunch Hour (Noon - 1 pm) PM Session (1 - 4)

	Day 1	Hr	# of Slides
AM	**CH01:** Introduction to Process Measurement	1	
	CH02: Principles of Transducer Operation	1	
	CH03: Basic Process Measurement Systems	1	
PM	**CH04:** Systems Standards and Instrument Calibration	1	
	CH05: Maintaining System Quality	1	
	Total Contact Hours	5	
	Certification Exam	2	

MSOE-PERD   www.msoe.edu/seminars   khalil@msoe.edu   Cell: +1-414-940-2232

Pressure Measurement

		Process Engineering						
Course #	Course Title	CEU	Hr	Days	Hands-On	Exam	Scheduled	$/Person
TPC 273	Pressure Measurement	0.5	5	1	✗	✓	✗	$250

Course Description:

This 5-hour 1-day course covers units of pressure and discusses Boyle's and Charles' laws to explain relationships among pressure, volume, and temperature. Describes sensor operation of manometers, bourdon tubes, diaphragms, and bellows. Explains the operation of potentiometric, capacitive, reluctive, servo, strain-gauge, and piezoelectric transducers. Describes devices used in low-pressure control. Discusses proper and safe methods for installing and servicing pressure instruments.

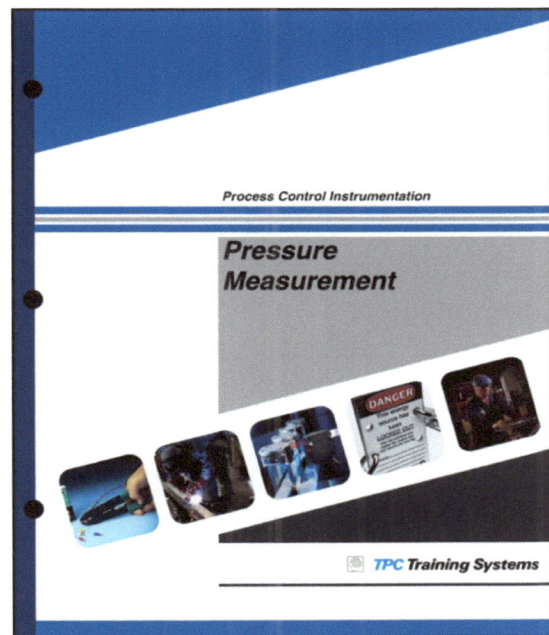

Course Agenda:

Course Outline/Agenda: AM Session (9-Noon)   Lunch Hour (Noon - 1 pm)   PM Session (1 - 4)

	Day 1	Hr	# of Slides
AM	**CH01:** Principles of Pressure in Liquids and Gases	1	
	CH02: Pressure Sensors	1	
	CH03: Pressure Transducers	1	
PM	**CH04:** Low-Pressure Measurement	1	
	CH05: Installation and Service	1	
	Total Contact Hours	5	
	Certification Exam	2	

MSOE UNIVERSITY

Force, Weight, and Motion Measurement

Process Engineering

Course #	Course Title	CEU	Hr	Days	Hands-On	Exam	Scheduled	$/Person
TPC 274	Force, Weight, and Motion Measurement	0.5	5	1	✗	✓	✗	$250

Course Description:

This 5-hour 1-day course covers force, stress, and strain and explains the operation of strain-gauge systems. Relates weight to mass and scales to balances. Explains the operation of load-cell scales. Describes belt-scale, nuclear-scale, and weigh feeder operation. Covers position measurements by means of proximity detection, air gauging, LVDT gauges, synchros, code disks, and other devices. Explains machine tool control and accelerometer operation. Describes the measurement of angular velocity and acceleration, vibration detection, and machinery balancing.

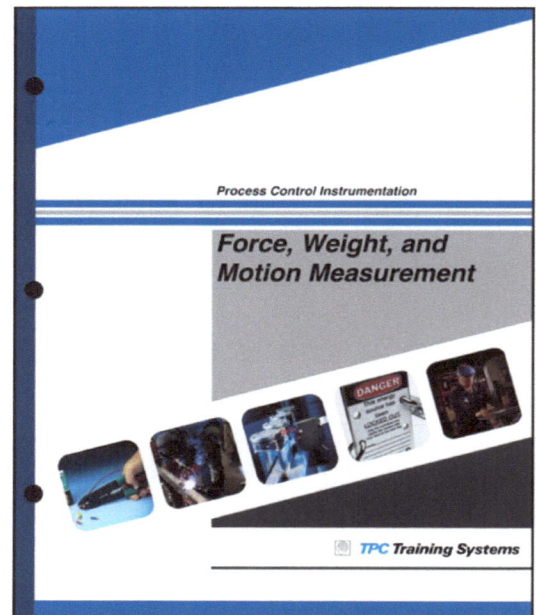

Course Agenda:

Course Outline/Agenda: AM Session (9-Noon)   Lunch Hour (Noon - 1 pm)   PM Session (1 - 4)

	Day 1	Hr	# of Slides
AM	**CH01:** Force, Stress, and Strain	1	
	CH02: Weight and Mass Measurement	1	
	CH03: Weighing Materials in Motion	1	
PM	**CH04:** Position Measurements	1	
	CH05: Acceleration, Vibration, and Shock	1	
	Total Contact Hours	5	
	Certification Exam	2	

Flow Measurement

Process Engineering

Course #	Course Title	CEU	Hr	Days	Hands-On	Exam	Scheduled	$/Person
TPC 275	Flow Measurement	1.0	10	2	✗	✓	✗	$500

Course Description:

This 10-hour 2-day seminar covers principles of fluid flow and how primary devices affect fluid flow. Describes flow measurement using several kinds of secondary devices. Discusses rotameters and other variable-area instruments. Explains how weirs, flumes, and other arrangements measure open-channel flow. Compares many kinds of positive-displacement meters and explains the operation of several kinds of turbine and magnetic flow meters. Describes less-common flow meters (including vortex-precession, mass flow, and ultrasonic devices) and instruments that meter the flow of solids. Provides guidelines for safe installation and maintenance of flow devices.

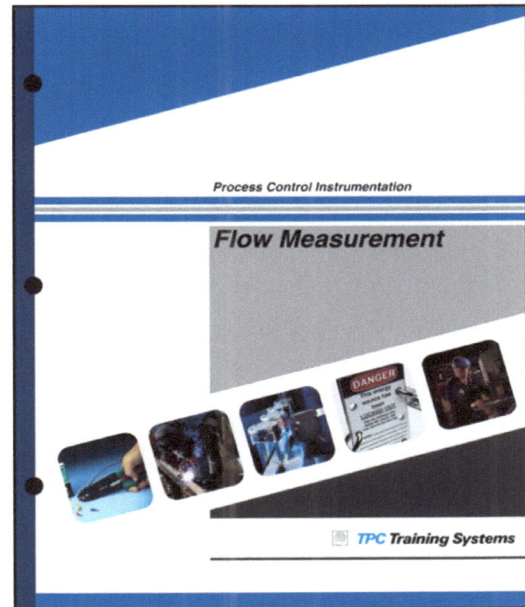

Process Control Instrumentation

Flow Measurement

TPC Training Systems

Course Agenda:

Course Outline/Agenda: AM Session (9-Noon) Lunch Hour (Noon - 1 pm) PM Session (1 - 4)		Hr	# of Slides
	Day 1	**Hr**	**# of Slides**
AM	**CH01:** Properties of Fluid Flow	1	
	CH02: Primary Measuring Devices	1	
	CH03: Secondary Measuring Devices	1	
PM	**CH04:** Variable-Area Instruments	1	
	CH05: Open-Channel Flow Devices	1	
	CH06: Positive-Displacement Meters	1	
	Day 2	**Hr**	
AM	**CH07:** Turbine and Magnetic Flow meters	1	
	CH08: Specialized Flow meters	1	
	CH09: Metering the Flow of Solid Particles	1	
PM	**CH10:** Installation and Maintenance of Flow Instruments	1	
	Total Contact Hours	10	
	Certification Exam	2	

Level Measurement

Course #	Course Title	CEU	Hr	Days	Hands-On	Exam	Scheduled	$/Person
	Process Engineering							
TPC 276	Level Measurement	0.5	5	1	✗	✓	✗	$250

Course Description:

This 5-hour 1-day course covers principles governing various methods of measuring level. Explains operation of conductive, capacitive, resistive, ultrasonic, and photoelectric devices. Compares the operation of several kinds of pressure-head instruments. Explains the measurement of solids by ultrasonic, microwave, radiation, and other methods. Discusses several special-application devices for both continuous and point level measurement.

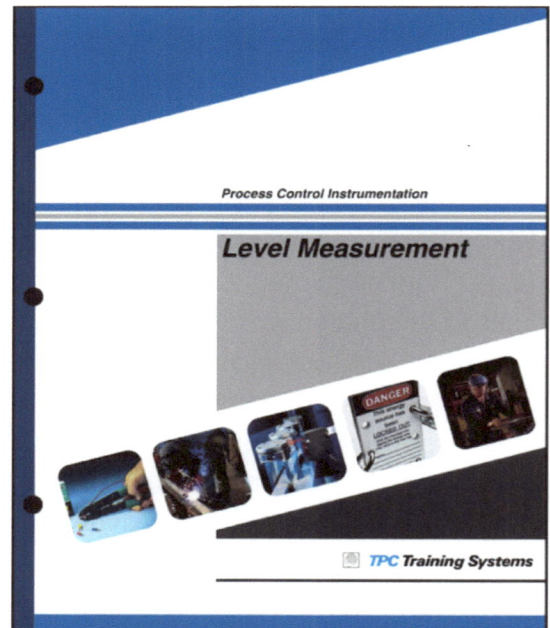

Course Agenda:

Course Outline/Agenda: AM Session (9-Noon) Lunch Hour (Noon - 1 pm) PM Session (1 - 4)

	Day 1	Hr	# of Slides
AM	**CH01:** Principles of Level Measurement	1	
	CH02: Electrical Instruments	1	
	CH03: Pressure Head Instruments	1	
PM	**CH04:** Solid Level Measurement	1	
	CH05: Other Level Measurement Instruments	1	
	Total Contact Hours	5	
	Certification Exam	2	

Working with Controllers

Process Engineering

Course #	Course Title	CEU	Hr	Days	Hands-On	Exam	Scheduled	$/Person
TPC 281	Working with Controllers	0.5	5	1	✗	✓	✗	$250

Course Description:

This 5-hour 1-day course covers the purposes and kinds of controllers and their relationship to other components in process control systems. Explains the concepts of current-, position-, and time-proportioning control. Compares the operation of several kinds of controllers. Describes the operation of proportional, integral, and derivative modes, and discusses tuning procedures for each. Discusses cascade, feedforward, ratio, and auctioneering control systems as well as other operations. Describes ways to eliminate or reduce controller problems.

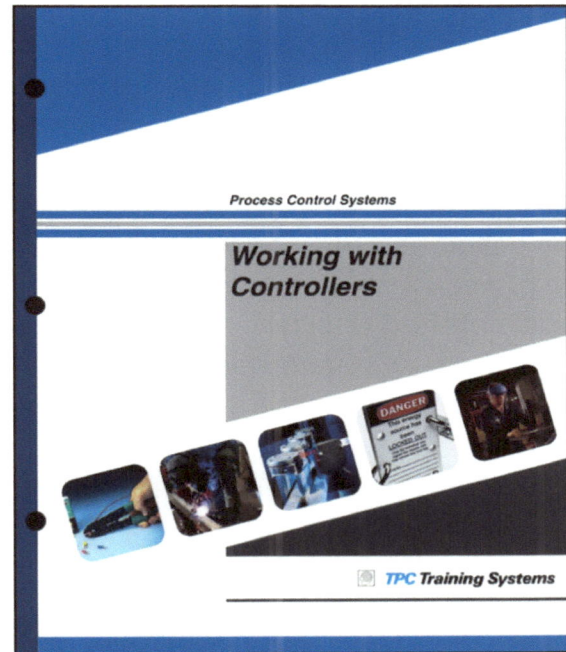

Course Agenda:

Course Outline/Agenda: AM Session (9-Noon) Lunch Hour (Noon - 1 pm) PM Session (1 - 4)			
	Day 1	Hr	# of Slides
AM	**CH01:** Introduction to Controls	1	
	CH02: Controller Operations	1	
	CH03: Controller Modes and Tuning	1	
PM	**CH04:** Special Controller Applications and Options	1	
	CH05: Maintaining Controller Systems	1	
	Total Contact Hours	5	
	Certification Exam	2	

How Control Loops Operate

		Process Engineering						
Course #	Course Title	CEU	Hr	Days	Hands-On	Exam	Scheduled	$/Person
TPC 282	How Control Loops Operate	0.5	5	1	✗	✗	✗	$250

Course Description:

This 5-hour 1-day course covers definition of control loop terms and characteristics. Includes specific examples of operation of control loops of many kinds. Discusses proportional, integral, and derivative modes in detail. Describes advanced control methods by means of four strategies with specific examples. Examines the effects of loop dynamics on system stability.

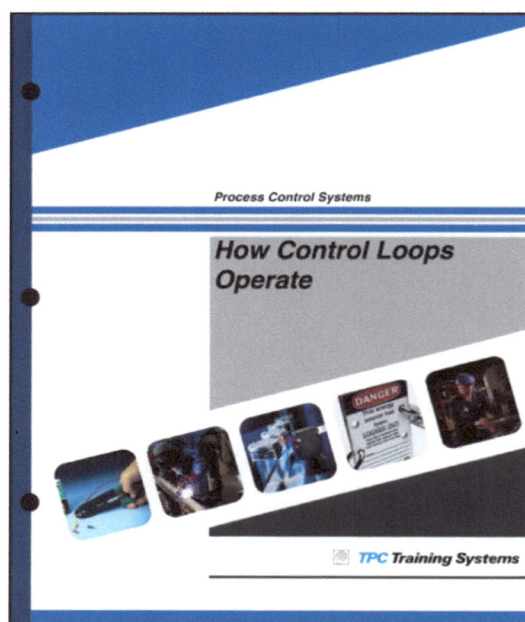

Course Agenda:

Course Outline/Agenda: AM Session (9-Noon)   Lunch Hour (Noon - 1 pm)   PM Session (1 - 4)

	Day 1	Hr	# of Slides
AM	**CH01:** Fundamentals of Control Loops	1	
	CH02: Control Loop Characteristics	1	
	CH03: Advanced Control Methods	1	
PM	**CH04:** Loop Dynamics	1	
	CH05: Loop Protection	1	
	Total Contact Hours	5	
	Certification Exam	2	

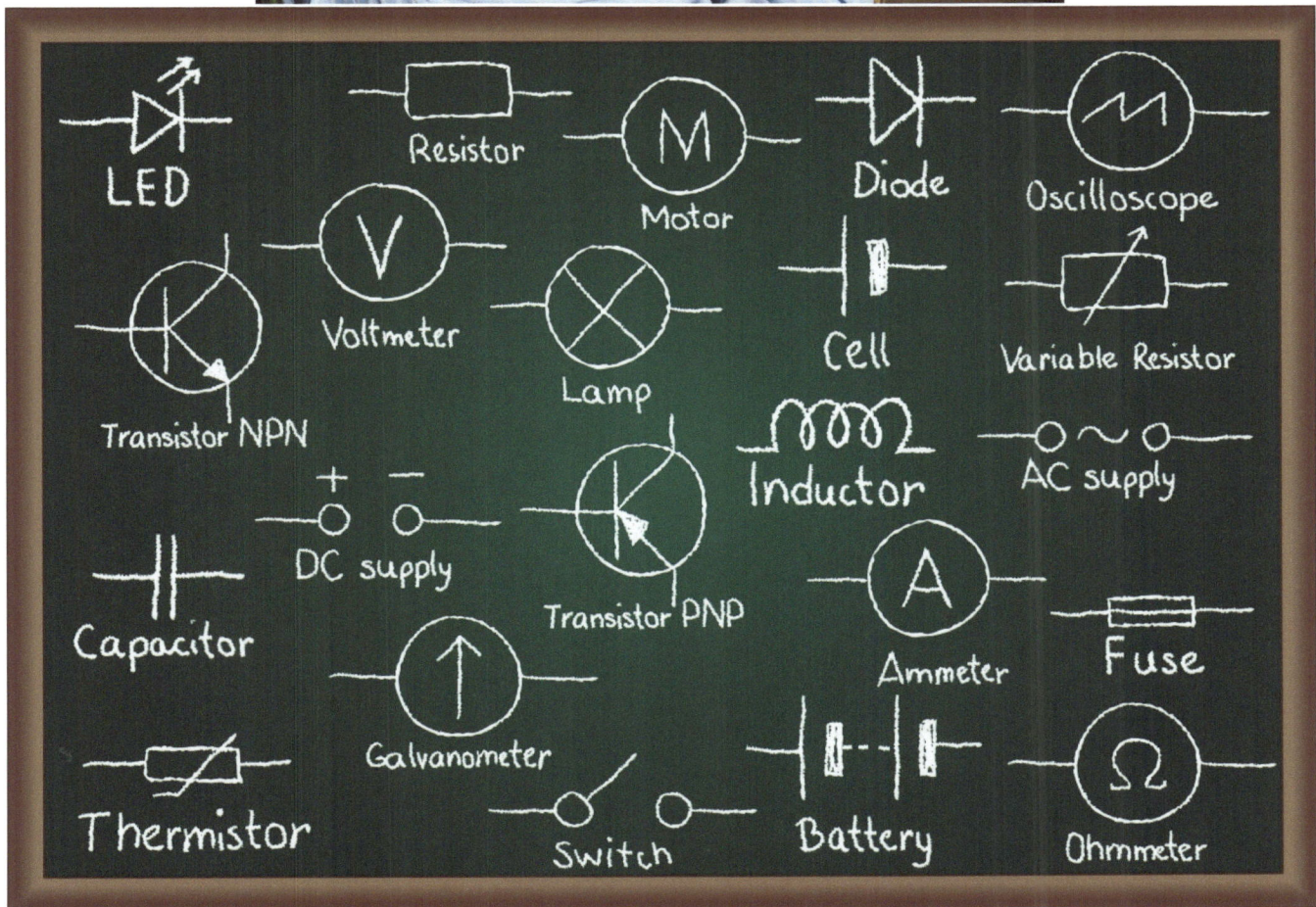

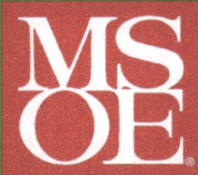

Designation Table

Condition	Des.	Clarification
Exam:	✓	Course contains certification exam to get certified
	✗	No certification exam.
Hands-On:	✓	Course contains hands-on labs.
	✗	Course conducted on theoretical base.
Scheduled:	✓	Course scheduled and registration is opened for public
	✗	Course is offered upon request at the customer-site or for public when the minimum enrollment number is reached.
	UD	Course is under development.

Customize Your Own Industrial Training.
Courses can be mobilized to your facility.
Courses in this sectors are non-scheduled courses offered only in customer-site.
If there is an interest, please contact Dr. Medhat Khalil directly.

Electrical Systems

Course #	Course Title	CEU	Hr	Days	Hands-On	Exam	Scheduled	$/Person
ATP08A	Electrical Principles and Practices I	2.1	21	4	✗	✓	✗	$1,050
ATP08B	Electrical Principles and Practices II	2.7	27	5	✗	✓	✗	$1,350
ATP09	AC/DC Principles	2.7	27	5	✗	✓	✗	$1,350
ATP10	Introduction to Programmable Logic Controls	2.7	27	5	✗	✓	✗	$1,350
ATP11	Motors	2.7	27	5	✗	✓	✗	$1,350
ATP12A	Electrical Motor Controls for Integrated Systems I	2.4	24	4	✗	✓	✗	$1,200
ATP12B	Electrical Motor Controls for Integrated Systems II	2.4	24	4	✗	✓	✗	$1,200
ATP13	Solid State Devices and Systems	2.7	27	5	✗	✓	✗	$1,350

Electrical Principles and Practices I

Electrical Systems

Course #	Course Title	CEU	Hr	Days	Hands-On	Exam	Scheduled	$/Person
ATP08A	Electrical Principles and Practices I	2.1	21	4	✗	✓	✗	$1,050

Course Description:

This 21-Hour 4-Days seminar is PART 1 of an introduction to electrical and electronic principles and practices and their uses in residential, commercial, and industrial applications. Chapters have been expanded to include greater coverage of personal protection and safety. A CD-ROM is included and contains information to supplement the textbook.

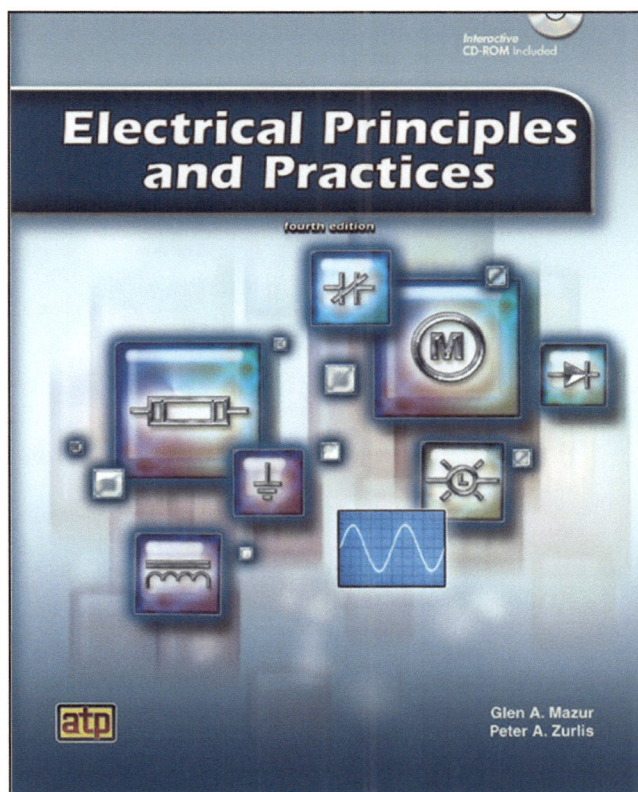

Course Agenda

Course Outline/Agenda: AM Session (9-Noon) Lunch Hour (Noon - 1 pm) PM Session (1 - 4)

	Day 1	Hr	# of Slides
AM	**CH01:** Electricity Principles	1	19
	CH02: Basic Quantities	2	31
PM	**CH03:** Ohm's Law and the Power Formula	1.5	20
	CH04: Safety	1.5	26
	Day 2	Hr	
	CH05: Math Principles	3	46
PM	**CH06:** Math Applications	1.5	22
	CH07: Numbering Systems and Codes	1.5	23
	Day 3	Hr	
AM	**CH08:** Meter Abbreviations and Displays	1	14
	CH09: Taking Standard Measurements	2	34
PM	**CH10:** Symbols and Pintreading	1.5	36
	CH11: Circuit Conductors Connections and Protection	1.5	38
	Day 4	Hr	
	CH12: Series Circuits	1	15
AM	**CH13:** Parallel Circuits	1	14
	CH14: Series Parallel Circuits	1	14
	Total	21	

Electrical Principles and Practices II

		Electrical Systems						
Course #	Course Title	CEU	Hr	Days	Hands-On	Exam	Scheduled	$/Person
ATP08B	Electrical Principles and Practices II	2.7	27	5	✘	✓	✘	$1,350

Course Description:

This 27-Hour 5-Day seminar is PART 2 of an introduction to electrical and electronic principles and practices and their uses in residential, commercial, and industrial applications. Chapters have been expanded to include greater coverage of personal protection and safety. A CD-ROM is included and contains information to supplement the textbook. X

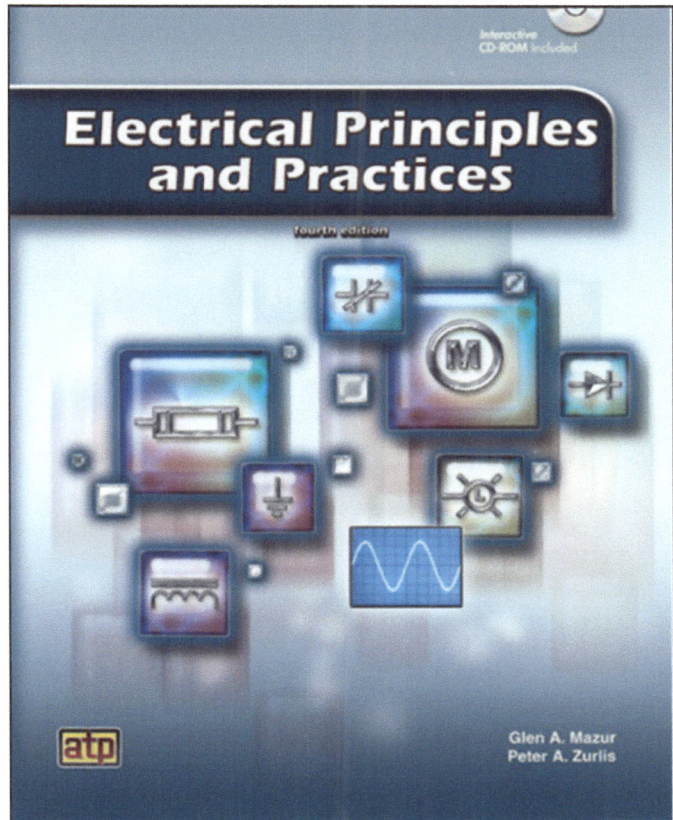

Course Agenda

Course Outline/Agenda: AM Session (9-Noon)   Lunch Hour (Noon - 1 pm)   PM Session (1 - 4)

		Day 1	Hr	# of Slides
AM		**CH15:** Transformers and Smart Grid Technology	3	38
PM		**CH16:** Electric Motors	1.5	27
		CH17: Resistance, Inductances and Capacitance	1.5	25
		Day 2	Hr	
AM		**CH18:** Circuit Requirements	3	40
PM		**CH19:** Residential Circuits	3	32
		Day 3	Hr	
AM		**CH20:** Commercial Circuits	3	47
PM		**CH21:** Industrial Circuits	3	41
		Day 4	Hr	
AM		**CH22:** Fluid Power Circuits	1.5	29
		CH23: Audio Systems	1.5	32
PM		**CH24:** Electrical Control Devices	1.5	29
		CH25: Digital Electrical Circuit	1.5	25
		Day 5	Hr	
AM		**Review and conclusion**	1	
		Certification Exam	2	
		Total		27

AC/DC Principles

Electrical Systems

Course #	Course Title	CEU	Hr	Days	Hands-On	Exam	Scheduled	$/Person
ATP09	AC/DC Principles	2.7	27	5	✗	✓	✗	$1,350

Course Description:

This 27-Hour 5-Days seminar shows learners how to apply basic laws and analysis techniques to introductory circuits as well as actual AC and DC circuit applications. Ohm's law, Kirchhoff's law, theories applied to basic circuits. Algebra and trigonometry are applied to aid in building fundamental mathematical skills. Step-by-step example problems follow all mathematical formulas. The seminar also includes an introduction to concepts of electricity, network analysis techniques, and vector diagrams and phase relationships, and concludes with chapters on resonance, three-phase AC, transformers, and AC motors. A CD-ROM is included and contains information to supplement the textbook.

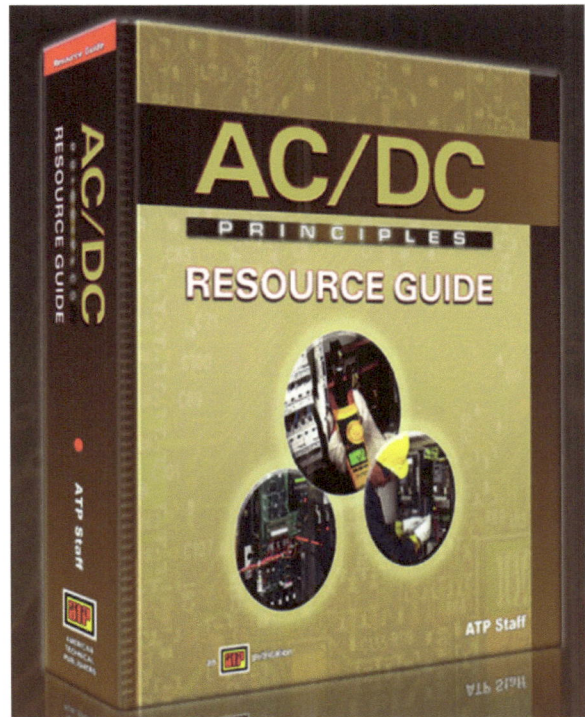

Course Agenda:

Course Outline/Agenda: AM Session (9-Noon)   Lunch Hour (Noon - 1 pm)   PM Session (1 - 4)

	Day 1	Hr	# of Slides
AM	**CH01:** Basic Concepts of Electricity	1.5	41
AM	**CH02:** Resistance	1.5	37
PM	**CH03:** Voltage Sources	1.5	42
PM	**CH04:** The Simple Circuit and Ohm's Law	1.5	31
	Day 2	**Hr**	
AM	**CH05:** DC Series Circuits	1	13
AM	**CH06:** DC Parallel Circuits	1	15
AM	**CH07:** DC Series/Parallel Circuits	1	12
PM	**CH08:** Complex Network Analysis Techniques	1	13
PM	**CH09:** Electromagnetism	1	24
PM	**CH10:** DC Circuit Inductance	1	18
	Day 3	**Hr**	
AM	**CH11:** DC Circuit Capacitance	1	19
AM	**CH12:** AC Fundamentals	1	16
AM	**CH13:** Vectors and Phase Relationships	1	17
PM	**CH14:** Resistive AC Circuits	1	13
PM	**CH15:** Inductive AC Circuits	1	20
PM	**CH16:** Capacitive AC Circuits	1	20
	Day 4	**Hr**	
AM	**CH17:** Inductive-Resistive-Capacitive Circuits	1	11
AM	**CH18:** Resonance	1	22
AM	**CH19:** Three-Phase AC	1	22
PM	**CH20:** Transformers	2	23
PM	**CH21:** AC Motors	1	15
	Day 5	**Hr**	
AM	**Review and conclusion**	1	
AM	**Certification Exam**	2	
	Total	27	

MSOE-PERD   www.msoe.edu/seminars   khalil@msoe.edu   Cell: +1-414-940-2232

Introduction to Programmable Logic Controls

		Electrical Systems						
Course #	Course Title	CEU	Hr	Days	Hands-On	Exam	Scheduled	$/Person
ATP10	Introduction to Programmable Logic Controls	2.7	27	5	✗	✓	✗	$1,350

Course Description:

This 27-Hour 5-Days seminar covers the fundamentals of installing, programming, and troubleshooting digital and analog PLCs. Introduction to Programmable Logic Controllers is a text/workbook that provides a solid foundation in PLC theory, installation, programming, operation, and troubleshooting. Many large, detailed drawings of commercial and industrial PLC systems are used to support the information in the textbook. Electrical Principles and Practices is an introduction to electrical and electronic principles and practices and their uses in residential, commercial, and industrial applications. Chapters have been expanded to include greater coverage of personal protection and safety. A CD-ROM is included and contains information to supplement the textbook.

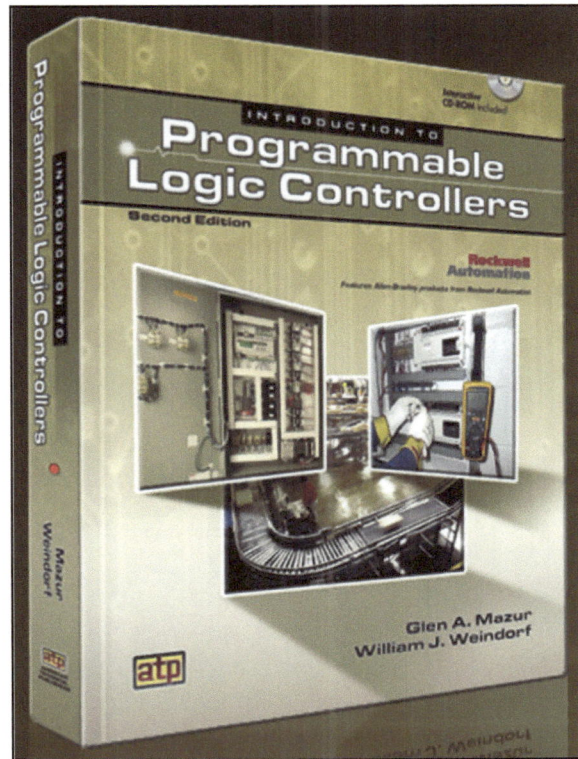

Electrical Systems Training

Course Agenda

Course Outline/Agenda: AM Session (9-Noon)   Lunch Hour (Noon - 1 pm)   PM Session (1 - 4)

	Day 1	Hr	# of Slides
AM	Pre-Test	1.5	0
AM	CH01: PLC and Electrical Safety	1.5	41
PM	CH02: Electrical Principles and PLC's	1.5	28
PM	CH03: Electrical Circuits and PLC's	1.5	27
	Day 2	Hr	
AM	CH04: PLC Hardware	1	19
AM	CH05: PLC Programming Instructions	1	18
AM	CH06: Programming PLC Timers and Controllers	1	17
PM	CH07: PLC and System Interfacing	1.5	31
PM	CH08: PLC Installations and Start Up	1.5	26
	Day 3	Hr	
AM	CH09: PLC and System Maintenance	1	13
AM	CH10: Troubleshooting Principles and Test Instruments	2	32
PM	CH11: Troubleshooting PLC Hardware	1.5	24
PM	CH12: Troubleshooting with PLC Software	1.5	16
	Day 4	Hr	
AM	CH13: Analog Principles	1	13
AM	CH14: Analog Device Installation and PLC Programming	2	23
	Application Studies	3	
	Day 5	Hr	
AM	Review and conclusion	1	
AM	Certification Exam	2	
	Total	27	

Motors

Electrical Systems

Course #	Course Title	CEU	Hr	Days	Hands-On	Exam	Scheduled	$/Person
ATP11	Motors	2.7	27	5	✗	✓	✗	$1,350

Course Description:

This 27-Hour 5-Days seminar provides a comprehensive overview of electrical theory and fundamental motor operating principles as they relate to installation and troubleshooting procedures. This full-color textbook includes the latest information on motor operating principles, starting, braking, and the mechanical aspects of installing and operating motors. Motors is designed to help the learner understand both fundamental and advanced concepts. Many different types of specialized motors are explained. Installation, maintenance, and troubleshooting are discussed in detail. Motors also presents correct safety procedures in compliance with the National Electrical Code® and NFPA 70E®. It can be used in a classroom learning situation, as a self-study textbook, or as a reference book on specialized motors applications. A CD-ROM is included and contains information to supplement the textbook.

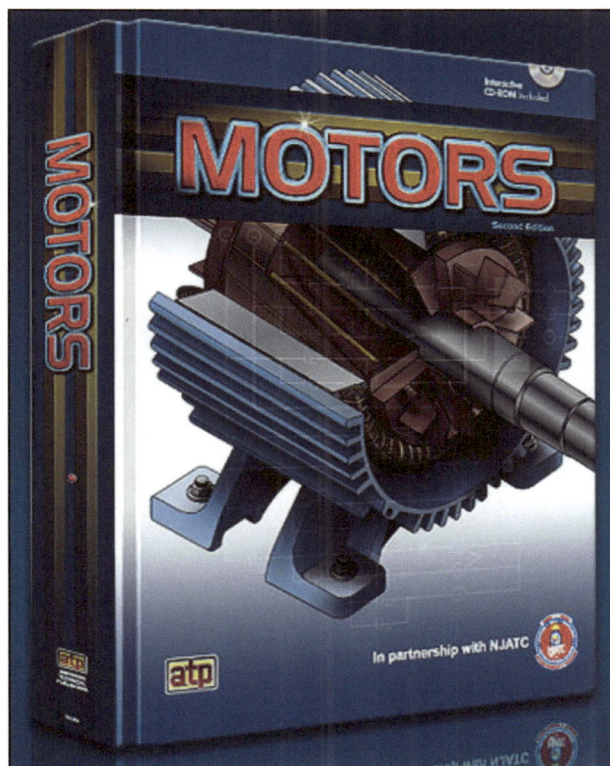

Course Agenda

Course Outline/Agenda: AM Session (9-Noon)   Lunch Hour (Noon - 1 pm)   PM Session (1 - 4)

	Day 1	Hr	# of Slides
AM	**CH01:** Magnetism and Induction	0.75	17
	CH02: Motor Construction and Nameplates	1.5	28
	CH03: AC Alternators	0.75	13
PM	**CH04:** Three-Phase Motors	2	31
	CH05: Squirrel-Cage Motors	1	6
	Day 2	**Hr**	
AM	**CH06:** Wound-Rotor Motors	0.5	12
	CH07: Synchronous Motors	1.5	25
	CH08: Single-Phase Motors	1	20
PM	**CH09:** Motor Protection	1.5	23
	CH10: DC Motors and Generators	1.5	25
	Day 3	**Hr**	
AM	**CH11:** Starting	1.5	22
	CH12: Braking	1	13
	CH13: Multispeed Motors	0.5	9
PM	**CH14:** Adjustable-Speed Drives	1.5	20
	CH15: Bearings	1.5	24
	Day 4	**Hr**	
AM	**CH16:** Drive Systems and Clutches		20
	CH17: Motor Alignment		28
	CH18: Troubleshooting Motors	2	38
	CH19: Special-Application Motors	1	10
	Day 5	**Hr**	
AM	**Review and conclusion**	1	
	Certification Exam	2	
		Total	27

MSOE-PERD   www.msoe.edu/seminars   khalil@msoe.edu   Cell: +1-414-940-2232

Electrical Motor Controls for Integrated Systems I

Electrical Systems

Course #	Course Title	CEU	Hr	Days	Hands-On	Exam	Scheduled	$/Person
ATP12A	Electrical Motor Controls for Integrated Systems I	2.4	24	4	✗	✓	✗	$1,200

Course Description:

This 21-Hour 4-Day seminar is PART 1 of an introduction to electrical motor controls for integrated systems. The seminar covers electrical, motor, and mechanical devices and their use in industrial control circuits. This seminar provides the architecture and content for acquiring the knowledge and skills required in an advanced manufacturing environment. In this fast-changing manufacturing environment, technicians must be competent in various aspects of mechanical, electrical, and fluid power systems for productivity and success. The textbook also serves as a practical resource for maintenance technicians responsible for production equipment and HVAC equipment. The textbook begins with basic electrical and motor theory, builds on circuit fundamentals, and reinforces comprehension through examples of industrial applications. Special emphasis is placed on the development of troubleshooting skills throughout the text. A CD-ROM is included and contains information to supplement the textbook.

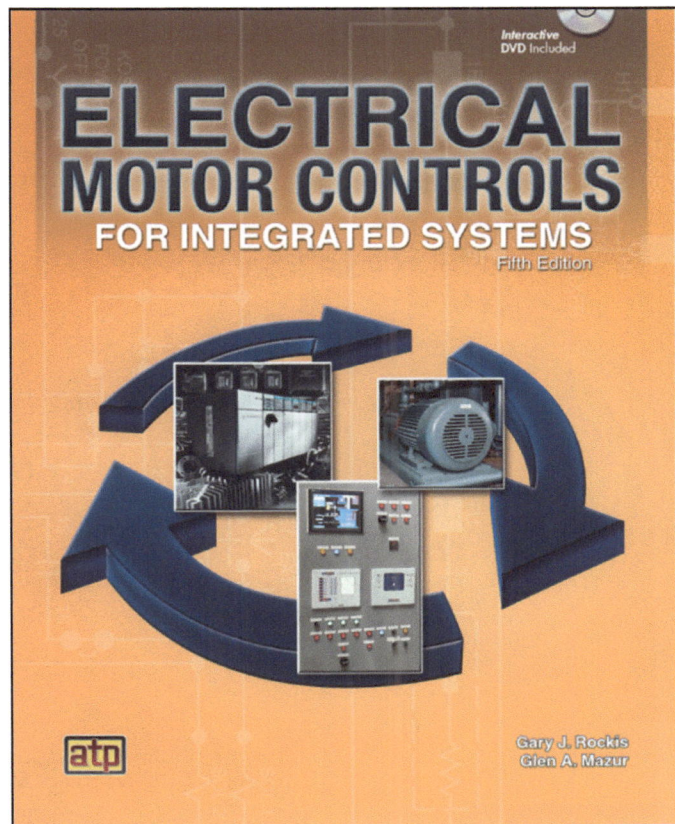

Course Agenda

Course Outline/Agenda: AM Session (9-Noon)   Lunch Hour (Noon - 1 pm)   PM Session (1 - 4)

	Day 1	Hr	# of Slides
AM	**CH01:** Electrical Quantities and Basic Circuits	2	35
	CH02: Symbols and Diagrams	1	23
PM	**CH03:** Test Instruments	1.5	29
	CH04: Electrical Safety	1.5	21
	Day 2	Hr	
AM	**CH05:** Control Logic	1.5	35
	CH06: Mechanical Input Control Devices	1.5	36
PM	**CH07:** Solenoids	1.5	19
	CH08: Electromechanical Relays	1.5	15
	Day 3	Hr	
AM	**CH09:** DC Generators	1	7
	CH10: AC Generators	1	13
	CH11: Transformers	1	13
PM	**CH12:** Contactors and Magnetic Motor Starters	3	43
	Day 4	Hr	
AM	**CH13:** DC Motors	1.5	31
	CH13: AC Motors	1.5	32
	CH15: Reversing Motors	1.5	31
	CH16: Timing and Counting Functions	1.5	41
	Total	24	

Electrical Motor Controls for Integrated Systems II

| | Electrical Systems | | | | | | | |
Course #	Course Title	CEU	Hr	Days	Hands-On	Exam	Scheduled	$/Person
ATP12B	Electrical Motor Controls for Integrated Systems II	2.4	24	4	✗	✓	✗	$1,200

Course Description:

This 21-Hour 4-Days seminar is PART 2 of an introduction to electrical motor controls for integrated systems. The seminar covers electrical, motor, and mechanical devices and their use in industrial control circuits. This seminar provides the architecture and content for acquiring the knowledge and skills required in an advanced manufacturing environment. In this fast-changing manufacturing environment, technicians must be competent in various aspects of mechanical, electrical, and fluid power systems for productivity and success. The textbook also serves as a practical resource for maintenance technicians responsible for production equipment and HVAC equipment. The textbook begins with basic electrical and motor theory, builds on circuit fundamentals, and reinforces comprehension through examples of industrial applications. Special emphasis is placed on the development of troubleshooting skills throughout the text. A CD-ROM is included and contains information to supplement the textbook.

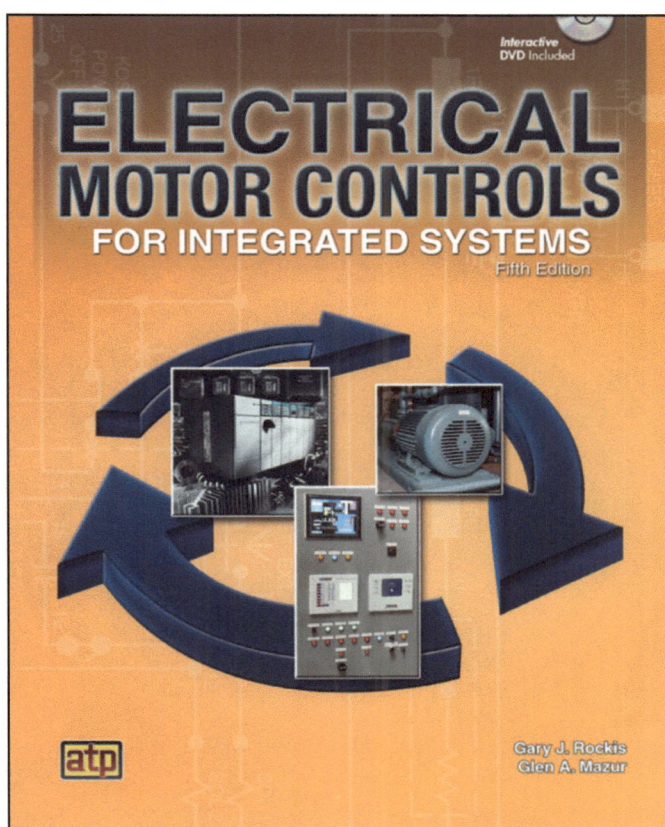

Course Agenda

Course Outline/Agenda: AM Session (9-Noon)   Lunch Hour (Noon - 1 pm)   PM Session (1 - 4)

	Day 1	Hr	# of Slides
AM	**CH17:** Motor Stopping Methods	1	12
	CH18: Motor Load, Torque, and Power Quality Requirements	1	17
	CH19: Reduced-Voltage Starting Circuits	1	19
PM	**CH20:** DC Power Sources	1.5	20
	CH21: Semiconductor Input Devices	1.5	33
	Day 2	**Hr**	
AM	**CH22:** Semiconductor Amplification and Switching	1.5	33
	CH23: Semiconductor Power Switching Devices	1.5	29
PM	**CH24:** Photoelectric Semiconductors, Fiber Optics, and Light-Based Applications	1.5	33
	CH25: Solid-State Relays and Starters	1.5	37
	Day 3	**Hr**	
AM + PM	**CH26:** Motor Drives	2	43
	CH27: Programmable Controllers	2	54
	CH28: Power Distribution and Smart Grid Systems	2	53
	Day 4	**Hr**	
AM	**CH29:** Preventive Maintenance Systems	2	31
	CH30: Predictive Maintenance	1	13
PM	**Review and conclusion**	1	
	Certification Exam	2	
	Total	**24**	

Solid State Devices and Systems

Electrical Systems

Course #	Course Title	CEU	Hr	Days	Hands-On	Exam	Scheduled	$/Person
ATP13	Solid State Devices and Systems	2.7	27	5	✗	✓	✗	$1,350

Course Description:

This 27-Hour 5-Day seminar presents a comprehensive overview of solid state devices and circuitry. This seminar is designed for electricians, students, and technicians who have a basic understanding of electricity. Component and circuit construction, operation, installation, and troubleshooting are emphasized and supported by detailed illustrations. Various practical applications are presented throughout the book as they relate to temperature, light, speed, and pressure control. Electron current flow is used throughout the book. Electron current flow is based on electron flow from negative to positive.

New and expanded topics include test instruments, printed circuit board construction, soldering and de-soldering, power sources and renewable energy, photonics, digital electronics, and solid state technology in programmable controllers. A CD-ROM is included and contains information to supplement the textbook.

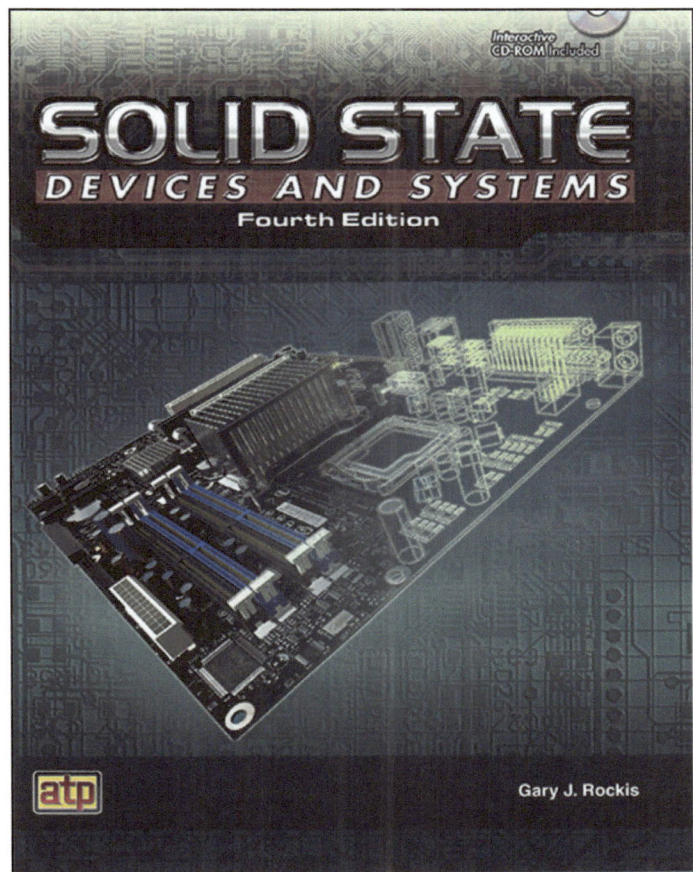

Course Agenda

Course Outline/Agenda: AM Session (9-Noon)   Lunch Hour (Noon - 1 pm)   PM Session (1 - 4)

	Day 1	Hr	# of Slides
AM	**CH01:** Symbols, Circuits & Safety	1.5	26
	CH02: Test Instruments	1.5	33
PM	**CH03:** Printed Circuit Board Construction & Troubleshooting	1.5	32
	CH04: Soldering and De-soldering	1.5	36
	Day 2	**Hr**	
AM	**CH05:** Diode Application and Trouble Shooting	1.5	33
	CH06: DC Power Supply Operation & Troubleshooting	1.5	38
PM	**CH07:** Power Sources and Renewable Energy	1.5	38
	CH08: Transducer Application and Troubleshooting	1.5	36
	Day 3	**Hr**	
AM	**CH09:** Bi-Polar Junction Transistor	1.5	40
	CH10: Transistors & Amplifiers	1.5	33
	CH11: JFET's, MOSFET's & IGBT's (Self Study)	0	36
	CH12: Silicone Controlled Rectifiers (Self Study)	0	33
PM	**CH13:** Triacs, Diacs and Unijunction Transistors	1.5	31
	CH14: Operational Amplifiers and 555 Timers	1.5	37
	Day 4	**Hr**	
AM	**CH15:** Photonics (Self Study)	0	63
	CH16: Digital Electronics Fundamentals (Self Study)	1.5	29
	CH17: Solid State Relays	1.5	37
	CH18: Engine Application and Selection	1.5	49
	CH19: Solid State Technology & Programmable Controls	1.5	28
	Day 5	**Hr**	
AM	**Review and conclusion**	1	
	Certification Exam	2	
	Total	**27**	

Work Safe

Be Safe

Designation Table

Condition	Des.	Clarification
Exam:	✓	Course contains certification exam to get certified
	✗	No certification exam.
Hands-On:	✓	Course contains hands-on labs.
	✗	Course conducted on theoretical base.
Scheduled:	✓	Course scheduled and registration is opened for public
	✗	Course is offered upon request at the customer-site or for public when the minimum enrollment number is reached.
	UD	Course is under development.

Customize Your Own Industrial Training.
Courses can be mobilized to your facility.
Courses in this sectors are non-scheduled courses offered only in customer-site.
If there is an interest, please contact Dr. Medhat Khalil directly.

Course #	Course Title	CEU	Hr	Days	Hands-On	Exam	Scheduled	$/Person
TPC109.1	Industrial Safety and Health	1.2	12	3	✗	✓	✗	$600
ATP14	Electrical Safety A Practical Guide to OSHA and NFPA 70E	1.5	15	3	✗	✓	✗	$750

Industrial Safety and Health

Course #	Course Title	CEU	Hr	Days	Hands-On	Exam	Scheduled	$/Person
TPC109.1	Industrial Safety and Health	1.2	12	3	✗	✓	✗	$600

Course Description:

This 12-hour 3-days seminar explains a safe workplace, discusses safety in various situations, personal protective equipment and fire safety. Includes expanded coverage of many health hazards. Covers ergonomics, environmental responsibility and importance of maintaining a safe work environment.

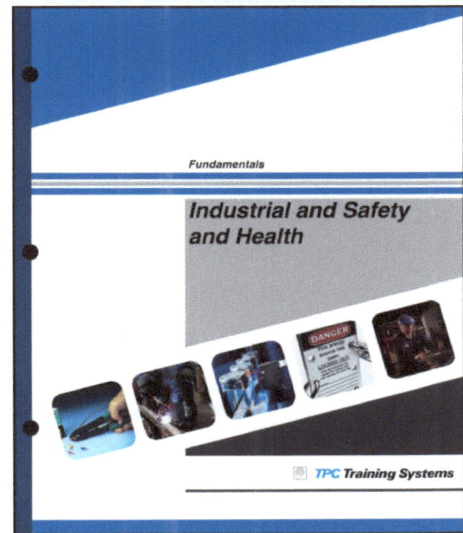

Course Agenda:

Course Outline/Agenda: AM Session (9-Noon) Lunch Hour (Noon - 1 pm) PM Session (1 - 4)

	Day 1	Hr	# of Slides
AM	CH01: Introduction to Safety and Health	1	
	CH02: Government Safety and Health Regulations	1	
	CH03: Personal Protective Equipment	1	
PM	CH04: Chemical Safety	1	
	CH05: Tool Safety	1	
	CH06: Material Handling	1	
	Day 2	**Hr**	**# of Slides**
AM	CH07: Working Safely with Machinery	1	
	CH08: Working Safely with Electricity	1	
	CH09: Electrical Equipment Safety	1	
PM	CH010: Fire Safety	1	
	CH011: Protecting Your Health 111	1	
	CH012: A Safe Work Environment	1	
	Total Contact Hours	12	
	Day 3	**Hr**	
AM	Certification Exam	2	

Electrical Safety A Practical Guide to OSHA and NFPA 70E

Course #	Course Title	CEU	Hr	Days	Hands-On	Exam	Scheduled	$/Person
ATP14	Electrical Safety A Practical Guide to OSHA and NFPA 70E	1.5	15	3	✗	✓	✗	$750

Course Description:

This 15-Hour 3-Day seminar is a comprehensive overview of electrical safety in the workplace. The seminar presents a practical guide to electrical safety as per OSHA and NFPA 70E®. The textbook features chapters on approach boundaries, working on energized circuits, establishing an electrically safe work environment, and choosing and inspecting personal protective equipment. The information provided helps learners understand how to reduce risk and avoid electrical hazards in the workplace while still being productive, which makes this textbook a valuable training tool for trainers, contractors, and electricians in the field. A CD-ROM is included and contains information to supplement the textbook.

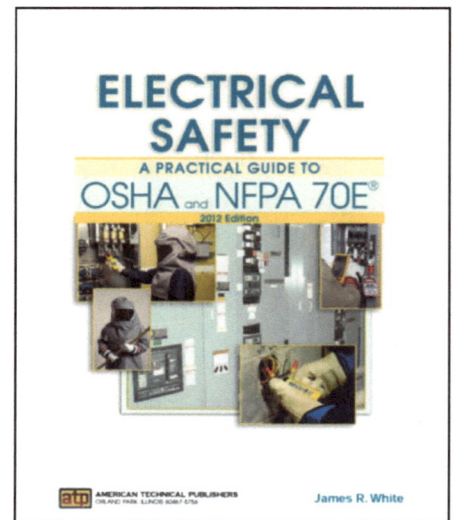

Course Agenda:

Course Outline/Agenda: AM Session (9-Noon) Lunch Hour (Noon - 1 pm) PM Session (1 - 4)		
Day 1	**Hr**	**# of Slides**
Pre-test	1.5	
AM **CH01:** Electrical Hazards and Basic Electrical Safety Concepts	1	12
CH02: Multi-Employer Worksites and Electrical Safety Programs	0.5	7
PM **CH03:** Training of Qualified and Unqualified Workers	0.75	8
CH04: Approach Boundaries for Shock and Arc Flash Hazards	0.75	15
CH05: Performing a Hazard/Risk Analysis	0.75	16
CH06: Establishing an Electrically Safe Work Condition	0.75	15
Day 2	**Hr**	
AM **CH07:** Working on Energized Conductors and Circuit Parts	1.5	33
CH08: Portable Electric Tools and Flexible Cords	1.5	20
PM **CH09:** Choosing and Inspecting Personal Protective Equipment	1.5	27
CH10: Guidelines for Common Electrical Tasks	1.5	24
Day 3	**Hr**	
AM **Review and conclusion**	1	
Certification Exam	2	
Total		**27**